1989

SIMULATIONS

A Handbook
for Teachers
and Trainers

SIMULATIONS

A Handbook for Teachers and Trainers

KEN JONES

**KOGAN
PAGE**

Copyright © 1980, 1987 Kenneth Jones
All rights reserved

First edition published in Great Britain
and the United States of America in 1980

This second edition first published in
Great Britain in paperback in 1987 by
Kogan Page Limited, 120 Pentonville Road,
London N1 9JN

British Library Cataloguing in Publication Data

Jones, Ken, *1923—*
    Simulations: a handbook for teachers and trainers —
    2nd ed. —
    1. Education — simulation methods
    I. Title
    371.3′97   LB1029.S53

    ISBN 1-85091-448-6

This second edition first published in the
United States of America in paperback in 1987 by
Nichols Publishing Company, PO Box 96,
New York NY10024

Library of Congress Cataloging-in-Publication Data

Jones, Ken, 1923—
    Simulations: a handbook for teachers.

    Bibliography: p.
    1. Education--Simulation methods--Handbooks, manuals,
etc.   I. Title.
LB1029.S53J66   1987   371.3′97   87-14813
ISBN 0-89397-286-X

Printed and bound in Great Britain
by Richard Clay Ltd,

Contents

371.397
J779⊥
2ⴹ

134,376

Preface

There have been two far-reaching changes in the educational field since the first edition of this book was published in 1980.

First, in Britain there has been a revolution in the curriculum which is attributed mainly to the advent of the General Certificate of Secondary Education. The GCSE represents a significant shift away from an instruction-dominated and fact-learning methodology towards greater emphasis on practical activities, oral skills, group work and interactive learning. The philosophy is based on the learner becoming an active participant in the learning process.

In the United States the educational system is still dominated by fact-learning methodologies, but there are signs that an influential American movement has begun in favour of interactive learning, and doubtless the example of the British educational revolution will not go unnoticed, and may set the tone of the debate.

The second important change is that there is now far greater awareness of what actually happens within simulations. The evidence sometimes provides surprising insights into other areas of education. For example, life skill courses which do not include simulations may actually increase anxieties about the world of work. There is also impressionistic evidence suggesting that in imaginative and open-ended simulations 11-year-olds are often more effective than the 14 to 15 year olds, and that black children are not only as good as whites but are better.

Both the GCSE revolution and the increased awareness of what happens in simulations emphasize the process of education rather than the product. It is a shift of emphasis from 'What do you know?' to 'How well can you cope?'. Thus, in a simulation which involves producing a television programme, the assessment should not be limited to the last five minutes of the event but cover the process — the false starts, the rewriting, the mistakes, the misunderstandings, the anxieties, the ethics, the attitudes, the emotions. These educational changes have required a substantial amount of rewriting and reshaping of this book. The main changes are as follows:

The last chapter of the first edition, 'The way ahead', has been dropped because it has been largely overtaken by events. The recommendations are no longer exhortations, they are reality.

The first chapter 'What's a simulation?' has been retained, but in a shorter and simplified form. It is also more critical of what I term 'hyphenated horrors' — simulation-games, simulation-exercises, simulation-dramas.

A new chapter 'Simulations as process' has been added to take account of the changes in educational philosophy and methodology since 1980. This provides a philosophical concept which can be used as a guide to action in the classroom.

The chapter 'Design' has been rewritten to include more specific advice for teachers (and pupils) on how to design simulations of their own.

The practical advice in 'Choosing simulations' and 'Using simulations' remains virtually intact, but with more options and advice in the section on de-briefing, and a new section has been added on computer-assisted simulations.

The final chapter on 'Assessment' has been substantially rewritten to take advantage of the greater wealth of detail which is now available concerning what people actually do and say in simulations. This chapter is particularly aimed at teachers who are uncertain about how to assess oral and behavioural skills in simulations. It contains transcripts from tapes to show how it is possible to award marks for oral and behavioural skills.

The book contains one significant change in terminology. The word 'controller' has been replaced by 'organizer' in order to reflect more accurately the non-intrusive and neutral aspects of running simulations, and consistent usage is desirable.

Quite a different problem is whether to use 'organizer' or 'teacher'. Here there is a desirable ambivalence because the actual thoughts are ambivalent. Sometimes they are thoughts about the event itself (organizer) while at the other times they can range over much wider educational issues (teacher). Teacher, throughout the book, is used in the widest sense to signify: schoolteachers, tutors, instructors, trainers facilitators and moderators.

Chapter 1

What's a simulation?

Essential characteristics

In this book 'simulation' refers to a classroom event which has two essential characteristics:

1. The participants have functional roles — survivor, journalist, judge, fashion designer, Prime Minister.
2. Sufficient information is provided on an issue or a problem — memos, maps, newspaper items, documents, materials — to enable the participants to function as professionals.

In the action part of a simulation there is no teacher. The participants must have autonomy, including the power and the authority to make mistakes. The survivors must be allowed the opportunity to die, the journalists must be allowed the opportunity to miss the deadline, the cabinet ministers must be given the freedom to fail to discuss the main item on the agenda. Thus, a simulation must be a non-taught event. If it is taught then it is not a simulation.

Teachers who are not used to simulations often have difficulty at first in resisting the temptation to interfere in the action in an effort to help the participants 'succeed'. Habits of help, guidance and instruction are not easily placed in cold storage. There is no policeman to stop a teacher walking into the cabinet office and sitting down next to the Prime Minister and saying 'How are you getting on? Do you understand what you have to do? Have you consulted with your colleagues yet?'

A basic reason for using simulations is that mistakes are both inevitable and desirable. It is experiential learning, not programmed learning, or a rehearsed event. Participants learn from their mistakes and want the opportunity to improve in the next simulation. The greater the disaster, the greater the learning. Any anxieties felt by the organizer that the participants 'may not get it right' are caused by a misunderstanding of the educational justification for simulations.

The provision of the key facts is essential. Simulations are not improvised drama, or episodic role play. They are not isolated events played out before an audience of fellow pupils/students. In a professional situation the participants must have the key facts, and not be asked to invent them. To say 'You are the customer returning a broken shoe, and you are the shop assistant' is not enough. Even though such role play may be functional rather than play-acting, this is not sufficient in itself to make it a simulation. Role play usually involves a high degree of participant authorship. In the above example the participants would probably invent key facts. The customer might take on the role of author and invent the circumstances in which the shoe broke, where, when, and what were the consequences. The shop assistant could take on the role of author in order to invent the policy of the shop towards disgruntled customers who return articles. There is nothing wrong with this; imagination and improvisation are fine. But the thoughts, motives, and tasks of a participant in such a role play are quite different from a simulation. This is not to say that participants in simulations never invent 'facts', but they should never be the key facts, otherwise it is either a badly designed simulation or, if there is a significant degree of participant authorship, not a simulation at all.

Non-essential characteristics

A simulation does not have to attempt to reproduce reality. In fact, the more an author tries to reproduce the real world with all its complexities and irrelevancies, the more unworkable the event becomes.

It is not just that reality is often unworkable in the classroom, reality is not always desirable from an educational point of view. Some simulations deliberately distort reality, or turn it on its head in order to provide a contrast to reality, or an optional reality. There are simulations set in imaginary countries, or in pre-history, or in the future, or in a fantasy world. These are not non-simulations or sub-standard simulations. Factors which help produce a good simulation include plausibility and consistency, not attempts to duplicate the real world.

As readers of academic literature on simulations will realize, this rejection of 'representing reality' utterly contradicts the academic definition. According to SAGSET (Society for the Advancement of Games and Simulations in Education and Training) and ISAGA (International Simulation and Gaming Association) a simulation is

A working representation of reality; it may be abstracted, simplified or accelerated model of the process. It allows students to explore systems where reality is too expensive, complex, dangerous, fast or slow.

This definition may be suitable for systems analysis, computer programs, model making and the like, but it does not reflect, in terminology or in content, the interactive human simulations referred to in this book. The terminology — 'representation', 'systems', 'models', 'students' — suggests not professional roles in non-taught events but a sort of programmed instruction geared to the fact-learning of things called models and systems. The jargon is indicative of the American approach to education, which tends to be based on step-by-step learning aimed at results which can be quantified by means of objective (and frequent) testing.

Another non-essential aspect of simulations is the view that simulations must have factual answers. Some do, some don't. A simulation is not the same as a puzzle, problem or textbook. Many of the most interesting simulations are open ended, and deal with values and opinions, emotions and attitudes. In such simulations the participants might well consider 'What questions should we ask?' rather than 'What answers should we give?'

It is not essential that a simulation should have clearly defined educational objectives. Some of the most famous and effective simulations are educationally ambiguous, and the process is usually far more important than the end product.

Nor is it essential that the action part of a good simulation should include effective learning. Experiential learning frequently occurs after rather than during an event. The action part of a simulation can include muddle and mistakes, and participants can become completely convinced that their own arguments and solutions are the best, and that other people are being unreasonable. Very often participants will believe that if something goes wrong it must be someone else's fault rather than their own. This does not make the event a non-simulation or a bad simulation. In a simulation it is not a crime to make mistakes, the crime is not to learn from the experience.

One advantage of simulations is that they have de-briefings, a period after the event for an appraisal of what really happened. Learning from experience must allow time for reflection on that experience, and the opportunity to try again. Instant enlightenment is no more an essential feature of simulations than it is of life outside the classroom.

Boundary lines

Simulations are non-taught events. They are best characterized not by their titles or by their aims, but by what actually goes on in the minds of the participants. Therefore, the thoughts and attitudes, and the emotions and behaviour, are the evidence for distinguishing between simulations and other interactive techniques.

Boundary lines cannot be drawn between techniques merely by inspecting the documents. For example, the documents in a case study could be identical to those in a simulation. The difference is that in a case study the students are looking at an event from the outside for the

11

purpose of gaining knowledge, whereas in a simulation they are on the inside, with the power and authority of professionals who are trying to cope with a developing situation. This is an extremely important difference.

It is not enough to inspect the format. To an observer who walks into a classroom and witnesses a few minutes of a discussion between participants in the role of government ministers there might be little if any indication whether the event was role play or a simulation. Boundary drawing requires evidence of motives and thoughts. If the participants were:

(a) imitating particular people or stereotypes, and/or
(b) inventing key facts about the subject matter, and/or
(c) thinking of themselves as students taking a role under the supervision of the teacher

then the activity would be in the field of drama in the case of (a), authorship in (b), and tuition in (c). But if, before the observer arrived, the participants had spent some time delving into the 'facts' about the issue, and were thinking professionally, then it would be a simulation.

Games share with simulations the essential criteria of autonomy. Once the action begins neither of them can be taught without changing the technique into something different — into a coaching session, a guided exercise, etc. The players in a game and the participants in a simulation are in charge within their particular environment, but only so long as they accept the conditions. The players in a game must accept the rules and must accept that the objective is to win. The participants in a simulation must accept full professional responsibilities. If these conditions are not met, then the activity requires another label.

Suppose, for example, that after a game of Monopoly one of the 'players' says, 'I did not try to win because I did not think it was right to knock down houses in order to build hotels.' This reveals that the person was not a player as defined or implied in the rules of the game, and for that person the event was not a game.

If, after a property development simulation, a participant says 'I was just trying to score points. I know I was not behaving in a way likely to serve the best interests of my company (tenants, local government) but I was having some fun' then that person was a player or funster. For that person (and perhaps, as a consequence of one person behaving unprofessionally, for some of the others) the event was not a simulation.

Terminology

It follows from the demarcation of territories that a simulation can go terribly wrong if the organizer uses inappropriate terminology. The wrong words lead to the wrong expectations, and the wrong expectations lead to the wrong behaviour.

Since the language of gaming and the theatre slip easily off the tongue it can be useful to draw up a list of words and phrases which could be used to introduce simulations and, more importantly, a list of words and phrases which could sabotage the event.

The lists below of 'appropriate' and 'inappropriate' words and phrases could serve as a starting point. There is a loose but not exact correspondence between pairs of items in each list.

Appropriate	*Inappropriate*
Simulation, event, participant, behaviour, action, organizer, facilitator, professional, function, functional, job, issues, ethics.	Game, drama, play, player, playing, act, acting, scoring, winning, losing, rules, teacher, pupil/student, puzzles, exercises.
A functional role, respecting people's opinions, professional considerations, part of the job, acting in the best interests of, running a simulation, ethical integrity, conscientious behaviour, appropriate conduct, doing one's best.	A playacting role, playing to win, point scoring, having fun and games, playing a part, acting it out, staging a performance, imitating and mimicking, doing the exercise, learning the lesson, finding the answer.

The 'inappropriate' list is not intended to reflect adversely on the techniques of games, informal drama and exercises. Games can be serious as well as fun, and drama can involve far more than just acting. The intention is simply to underline the fact that words carry with them all sorts of associational baggage, and until simulations become a familiar experience, both organizers and participants are easy victims of inadvertent sabotage by using the wrong labels. As will be argued in later chapters, a clear appreciation of boundary lines helps immeasurably in choosing suitable simulations, designing and running the events, evaluating the materials and assessing the behaviour, including both oral and written skills.

One other usage of the word simulation is worth noting. Like the words 'drama', 'exercise' and 'game' the word 'simulation' is often used as shorthand to mean the materials on which the event is based. In this sense one can say, 'We have plenty of plays (exercises, games, simulations) in the library.' There is usually no difficulty in discriminating between the two meanings within the context. Although the two meanings — event and materials — usually go arm in arm, this book does concentrate on simulations as events, and it is useful to try to get into the habit of visualizing simulations in terms of people rather than paper.

Hyphenated horrors

A description of the terminology is not complete without examining the hyphenated horrors — simulation-game, simulation-drama, simulation-exercise. I shall argue that such terms should be abandoned. In theory they are ambiguous, inconsistent and misleading, and in practice they result in mishmash events with consequences ranging from the mildly dissatisfying to the utterly disastrous.

Consider the ambiguities of 'simulation-game'. In one usage the hyphen means 'and' — 'simulations and games'. This meaning covers two distinct categories and is similar to saying eggs-bacon or chalk-cheese, and it occurs (as a '/') in the title of the SAGSET journal *Simulation/Games for Learning*, indicating that the journal is concerned with two categories — (a) simulations and (b) games.

However, in the academic literature, the hyphen (or dash, or space) usually signifies a separate and third category of event which is neither game nor simulation, but a combination of the two. The official definition used by SAGSET and ISAGA in their literature is:

> A SIMULATION GAME combines the features of a game (competition, co-operation, rules, players) with those of a simulation (incorporation of critical features of reality).

The absence of the hyphen seems to make no difference, since the meaning is a combination of two techniques, not a sub-category of one of them, as would be the case if the word 'simulation' were being used as an adjective making 'simulation game' a sub-category of games but not a sub-category of simulations. Thatcher, in the SAGSET publication *Introduction to Games and Simulations* (1986) helpfully supplies some examples. His list of simulation-games includes Monopoly, whereas his examples of games are Snakes and Ladders, Ludo, chess, Scrabble, football, tennis. This categorization is interesting, but it has certain difficulties. For one thing, it implies that the only games are abstract games and physical games/sports, and that board games about property development (car racing, banking, war) are not games at all.

Another difficulty is that the definition of 'simulation-game' contradicts the definition of 'simulation', which is supposedly a representation of reality. But it now becomes a very peculiar reality indeed since it must exclude co-operation and competition, features which it is assumed would need to be imported from the field of games in order to make a simulation-game. If the definition of 'simulation-game' is accepted then it becomes necessary to add exception clauses to the SAGSET/ISAGA definitions of games and simulations:

SAGSET/ISAGA	A simulation is a working representation of reality,
REWRITE	A simulation is a working representation of reality. except in cases which include co-operation, competition, rules, payoffs.
SAGSET/ISAGA	A game is played when one or more players compete or co-operate for payoffs according to a set of rules.
REWRITE	A game is played when one or more players compete or co-operate for payoffs according to a set of rules, except in cases which incorporate critical features of reality.

These muddled concepts arise because academics prefer to look at packages of materials, not events, thus overlooking significant differences.

From this a more fundamental question arises – are any definitions desirable? Why not rely on descriptions alone? This is recommended by Wittgenstein in his *Philosophical Investigations* (1968):

> Consider for example the proceedings that we call 'games'. I mean board games, card games, ball games, Olympic games, and so on. What is common to them all? Don't say 'There must be something in common, or they would not be called 'games' but look and see whether there is anything in common to all. For if you look at them you will not see something that is common to all, but similarities, relationships, and a whole series of them at that.... How should we explain to someone what a game is? I imagine that we should describe games to him, and we might add: 'This and similar things are called 'games'.'

Wittgenstein's insistence on the need to look and see is in direct contrast to the usual charts, boxes and flow diagrams which clutter a great deal of the literature on simulations, and of academic writing in general. In simulations the consequences are not academic. Take a practical example of classroom consequences. Suppose that an author says, 'In this farming simulation I have used a dice to determine the weather. A dice is a gaming element. Therefore, this is a simulation-game.'

The conclusion is incorrect, and the mistake arises because of a failure to look at the event from the point of view of the participants. For them it makes no difference at all whether a dice is rolled or whether the organizer's notes say, 'At the beginning of the first day tell the farmers that the weather is sunny and at the beginning of the second day explain that it is stormy.' The dice is not being used as a gaming element, it is being used as a mechanism for randomized authorship. One could imagine all the elements in the farming simulation being decided by dice – the available crops and livestock, the area of land, etc. — but this would not make it less of a simulation and more of a game, it would merely dispense with the author. The event is not a simulation-game, and to call it one is to invite the participants to treat it as a game rather than treat it professionally.

Unfortunately, genuine simulation-games do exist as mishmash events, with inherent contradictions. Some authors, especially Americans, often build into the event inappropriate point-scoring mechanisms either for gaming reasons or for the purpose of educational assessment. In a simulation about an assembly it is appropriate to count votes, but it is not appropriate to award a hundred marks for people who introduce successful resolutions, fifty marks for successful amendments, and ten marks for every minute a person speaks. Are the participants supposed to behave as parliamentarians or as gamesters?

Genuine simulation-exercises are also genuine muddles. It is quite normal and proper for exercises to be evaluated and marked, both externally and internally. But record sheets can be highly inappropriate in a simulation. I was once asked by a publisher if I would incorporate into a parliamentary simulation a form with two columns headed 'Points we agree on', 'Points we disagree on', which would be given to each Member of Parliament to carry around and fill in. As well as being implausible, it would have introduced an irreconcilable conflict between the thoughts of and behaviour of students and the thoughts and behaviour of parliamentarians.

Similar conflicts occur in simulation-dramas between professional conduct and acting. The instructions in a simulation-drama about business could say 'Participants must try to behave in the best interests of their company' and the roles cards say 'You are a friendly supervisor', 'You are a disgruntled sales representative'. Depending on circumstances, it could be in the best interests of the company for the supervisor to be non-friendly or the sales representative to be non-disgruntled. It is not unusual for a political simulation to have a role card saying something on the lines of 'You are a rebel leader and you are angry because...'. This requires the acting of anger, and suggests to the participant that the behaviour should be not only highly emotional but also stupid. It deprives the participant of the opportunity of behaving in other ways which may be more suitable in the circumstances. It need take only one participant to start ham acting for others to follow suit and a situation can arise where half the group are trying to take it seriously while the other half are mimicking and playacting.

It is not only the design which can cause muddled expectations, but also the classroom context. If the setting is a drama class, then the expectations can be of drama if the organizer does not make it clear that acting is forbidden and that professional behaviour is required.

On some occasions these hyphenated events can be emotionally intolerable. This is not altogether surprising since the situation is similar to Pavlov's experiments with dogs. They first learned to respond differently to two signs, say a circle and a square, and then the experimenter gradually made the signs more like each other. The unfortunate dogs tended to have the equivalent of a nervous breakdown.

For the organizer, the experience of running a hyphenated event can be unrewarding and even shattering. The situation is made worse because both the organizers of mishmash events and the participants themselves are usually unaware of the source of the sabotage. The muddle conceals the cause. Because the cause is concealed no appropriate remedy suggests itself. Consequently, the disaster is either sufficiently large to deter the organizing of similar events, or minor enough to permit other mishmash events, in which case a similar or worse muddle can unwittingly occur.

Definitions and descriptions

The main argument in this chapter is that muddled classroom events are caused by muddled concepts, and that muddled concepts are caused by three factors:

1. misleading terminology
2. the pursuit of definitions rather than descriptions, and
3. concentrating on packages of materials and objectives rather than on what actually occurs during a simulation.

The fact is that 'simulation' is not an appropriate label. It carries with it the unfortunate dictionary associations of imitation, mimicry, unreality. Ideally there should have been a new word for this unique classroom event, but it is ten or twenty years too late to change to something more suitable. Consequently, it is important to realize that 'simulation' is a technical word with a variety of possible meanings. It can legitimately be used in relation to a map, or to television graphics of a spacecraft, or to a clockwork mouse.

In the context of this book 'simulation' refers to untaught events in which the participants have roles and are required to accept the responsibilities and duties of professionals.

The above sentence could serve as a definition. In the first edition of this book I argued against definitions and in favour of descriptions, but said

> If a short definition really is necessary, perhaps it might be: 'Simulations are reality.' From the teacher's point of view at least it errs on the right side, the side of function...

17

Subsequently I offered two other definitions:

> A simulation is reality of function in a simulated and structured environment. (Jones, 1985a)

> A simulation in education is an untaught event in which sufficient information is provided to allow the participants to achieve reality of function in a simulated environment. (Jones, 1985a)

These definitions try to rescue 'simulation' from the dictionary-bound definitions given by ISAGA, SAGSET and most academic writers on the subject. But they suffer from the same basic inadequacy of all definitions, they do no more than build pigeonholes. They do not describe the pigeons.

The sort of simulations described in this book are events. They are not categories or representations. Basically they are not even about specific subject areas, they are about human beings, human values, human interaction. One might say to a human being 'How do you describe yourself?', but not 'How do you define yourself?' People who are unable to define themselves are not less human or less able to talk meaningfully about themselves. As Wittgenstein said in *Philosophical Investigations:*

> When I give the description 'The ground was quite covered with plants' — do you want to say I don't know what I am talking about until I can give a definition of a plant?

This book will concentrate on descriptions and allow simulations to speak for themselves.

Chapter 2

Simulations as process

War and business

The first organized use of simulations is usually attributed to the
Prussian army in the nineteenth century. It occured for behavioural
reasons. The Prussians had been dissatisfied with the recruitment of
officers, and decided that the interview and the pen and paper tests
were not enough. Consequently, they devised simulations to test
behaviour. Instead of asking 'How would you cope with situation X?'
the idea was to place the person in that situation, as far as was practical
and desirable, and see what happened.

The idea was later taken up by the British army. All sorts of behavioural
situations were devised in which the candidates had roles — officer,
survivor, agent, engineer. They were tests which revealed varying
degrees of ingenuity, co-operation, leadership, courage and other
aspects of military life. Some were non-interactive simulations in which
individual participants had to tackle a situation without help or
co-operation, but most were interactive simulations involving teams.
The simulation technique was also widely introduced into army
training.

The differences between these simulations and written tests affected
not only the participants but also the assessors. For example, the
assessors of a test paper usually wait until the end of the test and then
check the result. But with simulations the observation covers the whole
of the activities. Behaviour is very much a matter of process rather
than product, and assessors become very involved in behavioural tests
because they are looking at living people, not lifeless paper.

In the United States after the outbreak of the Second World War the
military were faced with the problem of how to recruit spies. The
original plan involved the recruitment of criminals on the grounds that
dirty work required criminal experience. This policy of recruitment
resulted in several disasters in the field, and a change in recruitment
techniques was deemed desirable. One option which was considered
was to advertise for people willing to undertake dangerous work, but

this was rejected on the grounds that it would bring in a flood of macho posturing cranks and psychopaths. At that time an American officer returned from Sandhurst military college and described the British methods of simulations. This was taken up by the Americans. Likely officers and men were seconded to a test centre organised by the OSS, the Office of Strategic Services, for recruitment tests lasting several days. After a good deal of trial and error, a system was evolved whereby each candidate had to adopt a cover story — name, birthplace, schooling, career, etc. — which were not his own, but related to things he knew about. All were instructed that they must stick to their cover stories at all times, except in condition 'X' which would be announced formally by one of the staff. At various times, particularly after failing a test, individual candidates would be interviewed in a relaxed atmosphere by a member of staff who would say something on the lines of 'Don't worry about it, few people get through that test, but I suppose you have had tough experiences in the past.' At this point some candidates blurted out their true experiences and revealed their actual background, thus breaking cover.

One particularly frustrating task was to take on the role of a supervisor who had to erect a construction by giving directions to two 'helpers'. The 'helpers' were members of staff who would obey orders to place the bits and pieces in position, but if the instructions were ambiguous they would do the wrong thing, and at times they would stop and offer 'helpful' suggestions. Sometimes a candidate would strike the helpers and drop out of the course there and then. (OSS 1948)

Subsequently, the OSS experiences were used as a basis for simulations in business and industry in the United States. As was pointed out, choosing a manager is no different in principle from choosing a spy. All that is needed is to see what sort of behaviour is desirable and undesirable in the job, and then devise appropriate simulations. These business assessment procedures were generally known as assessment centres, although sometimes there was no centre as such, and the assessment was carried out at whatever location was the most suitable. (Moses 1977)

In recent years there has been an intersecting of army and business simulations. At one time all simulations in the army were military action simulations, but recently management simulations have been introduced in Britain, West Germany and some other countries. Army officers participate in simulations which deal with such matters as the design and procurement of weapons, and participants may take the role of officials in the Ministry of Defence, commercial companies which manufacture weapons, and so on. The reason for this use of management simulations is military awareness of the growing importance of logistics.

Although the word 'game' is used from time to time in both the armed forces and in business it may now be on the decline. There are operations rooms and assessment centres rather than games rooms and games centres. The usual words are exercise, simulation, test, assessment. With the armed forces the word 'game' is usually avoided. In the British army there are what is known as TEWTs, Tactical Exercises Without Troops, not Tactical Games Without Troops. And when one country informs another that it intends to hold military manoeuvres near its borders, the phrase used is military exercises, not military games.

In business, the label 'games' is still popular among top management, and the reason for this is far from clear. Perhaps managers like the idea of playing games, or maybe the word signifies status, since the terminology for lower ranking staff is usually 'exercise' or 'simulation'. The phrase 'management game' is fairly widely used, and, perhaps significantly, there is no equivalent phrase for the rest of the staff. There are signs that gaming terminology, even among managers, is being replaced by more appropriate professional labels. However, the phrases 'war games' and 'business games' are still popular in the media. They are the terminology of the headline writers and the television and film makers. Many authors and simulation designers also are attracted to the word 'game', perhaps because it indicates a 'fun' element, and they probably believe, perhaps mistakenly, that it helps sell the product.

The argument is not that genuine business games and war games do not exist. They do, and in plentiful numbers, as can be seen by walking into a games shop. They involve players, the aim is to win, and the object of participation is enjoyment. A characteristic is that they can be played many times, whereas it is most unusual for a specific simulation (or exercise, or informal drama) to be rerun with the same group.

War and business are useful examples of areas in which simulations have been developed, but simulations are also widely used in recruitment and training for all jobs which require behavioural skills. They are an established part of police training (crime and accident simulations), air transport (flight deck simulators), the medical profession (simulated injuries), the law (mock trials). The terminology of gaming tends to be avoided at the professional level. Moreover, the aims of the simulations are related mainly to the process rather than the end results, so the observational data includes all the muddles, misunderstandings and false starts. The first thoughts as well as the second thoughts are part of the evidence.

Education and the academics

Education would seem to be the only major area of simulation usage where attention tends to be focused on the results rather than the process.

The clearest example of this occurs in the United States where the educational system is geared to results, far more so than in most other countries. This arises partly because of the way in which education is funded in the United States. High importance is attached to the end product in order to merit grants and other financial benefits. From this flows a series of consequences.

A major consequence is that the end product tends to be clearly defined and measurable so that success and failure can be reflected in statistics. Numbers are regarded as being more impressively objective than words, descriptions and opinions. Since statistics of measurable achievements have such an important status it follows that priority must be given to tests which reveal easily quantifiable results. And this in turn leads to emphasis on testing the recall of facts, rather than the response to open questions. The consequence of concentrating on examinations with closed questions leads to the encouragement of didactic teaching and programmed learning. Since the justification for the American system is the end result it follows that it is essential to specify precisely the nature of the task, for without clear objectives there can be no way of telling whether the right results have been achieved. Once the task has been defined the system requires that pupils and students should be tested, usually frequently tested, to make sure that they are on course. Given the nature of this system it is not surprising that the favoured method of instruction is step-by-step learning, and that the dominant educational psychology is behaviourist with emphasis on the frequent reinforcement of correct responses.

This step-by-step method of advancing towards precisely defined objectives influences American simulation design. For example, Richard Duke of the University of Michigan, advocates nine detailed and sequential steps in design (1979). Duke, who always uses the word 'game' calls his first step 'Specification for Game Design'. He says

> Game architects need a blueprint composed of carefully delineated, detailed game specifications. At the outset, they need to conform to a plan, providing a clear, concise picture of the product to be created.

Although American academics use the word 'game' far more frequently than is the case in Britain and Europe, it is rare to find any justification given for the usage. If confronted with the question, American academics charmingly acknowledge that they do not really mean 'game' but say that the label helps to convey the motivational element of the activity. However, this does not explain why the Americans rather than the Europeans misappropriate the label 'game'. Perhaps a more plausible explanation is that games have end results which are quantified into scores, thus giving the label an academic respectability in an educational curriculum dominated by objectives. Also, American society is highly competitive, and since competition is the nature of games the misapplication of the label 'game' is likely to be less obvious and less objectionable than in other countries.

In Britain and most other European countries, the competitive, scoring, and 'fun' aspects of games are likely to deter teachers from introducing such activities into the classroom. Consequently, the fact that the three major societies (International Simulation and Gaming Association, Society for the Advancement of Games and Simulations in Education and Training, North American Simulation and Gaming Association) link 'games' and 'simulations' in their titles could well be regarded as something of an own goal. Objectively there is no reason why each technique should not have its own society, and if pairing is desired then there are more appropriate educational partners for simulations — exercises, or role play, or case studies.

The comparisons between the 'product' and 'process' concepts of the curriculum in this chapter owe a good deal to Stenhouse's *An Introduction to Curriculum Research and Development* (1975). However, as Stenhouse pointed out, the situation is not as clear cut as might be implied from the description of the two philosophies. For example, there exists in the United States a strong school of criticism directed against the 'product' model. And within American literature on games and simulations there are frequent attacks on the consequences of giving such overwhelming priority to the pursuit of measurable objectives. For example, Greenblat (1981) argues forcibly that learners should be active, should be free from a dependence on authority, and should be allowed to reason for themselves.

There are some signs that this American movement in favour of learner autonomy and interactive learning is gathering strength. An example of this is the recent expansionist activities of the North American Simulation and Gaming Association (NASAGA) which has started to collect and publish imaginative simulations. The quality of human interaction in some of these events appears to be considerable. A brochure on these simulations can be obtained from The National Gaming Center at the University of North Carolina, 1 University Heights, Asheville NC 28804-3299.

The English educational revolution

Critics of a curriculum based on judging education by its end product have frequently pointed out that clearly defined objectives can often lead to unexpected events being overlooked, and that much of what is valuable in education is not open to objective measurement. Such criticisms have developed to such an extent in Britain during recent years that it is now being generally recognized that an educational revolution is taking place. And, unusually for revolutions, it is not against the establishment, it is being led by the establishment.

A harbinger was the Communicative Movement of the 1970s, a mini-revolution in the field of teaching English as a foreign language, based on the premiss that languages should not be taught for their own sake but in order that they may be used, and that language in use meant language as communication. Whereas the traditional didactic teaching of vocabulary, grammer and pronunciation had concentrated on fact learning, the Communicative Movement shifted the emphasis to oral and behavioural skills. What mattered for the innovators was the communication of meaning rather than the acquisition of grammatical accuracy. In its philosophy, psychology and methodology the Communicative Movement formed a creative whole, changing the attitudes and practices of teachers in many countries of the world. It was concerned with processes rather than products, active learning rather than passive, and interaction between students rather than a purely teacher-dominated approach.

The revolution which is going on at the moment in schools in Britain has no title, but as one of the main thrusts comes from the teaching of English in schools and colleges, the English educational revolution can serve as a temporary label.

The main architects were Her Majesty's Inspectors. The Government asked the Inspectorate to merge two secondary school examinations, both frequently criticized on various counts, mainly because they could be passed by swotting up the facts without really understanding the subjects. The Inspectors merged the two examinations and, in effect, hijacked them. They produced an examination which was different in kind, based on their own educational philosophy which they had been disseminating for many years piecemeal in countless reports and school inspections without making any really substantial inroads into educational practice. The new examination, the GCSE (General Certificate of Secondary Education) is based on learning by doing, not just analysing. Unlike the earlier exhortations, this is now a philosophy with teeth.

The philosophy is exemplified in Britain's primary schools where children tend to be treated as professionals. They build, construct, design, experiment. They look after the Wendy house, create models in the sandpits, decide how to dress the doll, and learn from experience by doing. Moreover, they are usually in charge of their own experiments. This acceptance of adult responsibility is a factor often overlooked, even by advocates of the British system of primary education. Of course, teachers help and instruct, but basically the children are used to working in groups and undertaking joint tasks with only the minimum of supervision. They are, in effect, professional designers, scientists and artists. Their decisions may be amateurish but their motives are professional.

The impact of this revolution has only just started, and it is already spreading downwards from GCSE level through the secondary schools and upwards into further and higher education.

As well as being more practical — a key word — the examination offers teachers the chance of continuous assessment of courses as an alternative to a final test. And because of the practical nature of the new methodology, more attention is given to oral work and group work. For the first time in Britain, an oral examination is compulsory in English. This is of enormous significance. The authority of compulsory examinations shapes the curriculum.

Interestingly enough, the London and East Anglian Board includes a group simulation as a set piece in its oral examination. The candidates are divided into committees of five, each member in turn taking the chair and presenting a theme followed by questions from the other members of the committee. The same number of marks can be awarded for the skills of asking/answering questions as for the individual presentations.

Naturally, the Inspectorate were not alone in their educational philosophy. The Government's Assessment of Performance Unit had for many years made surveys of children at the ages of 11 and 15 using small interactive exercises and simulations designed to test oral skills. Various initiatives and courses, particularly the TVEI (Technical and Vocational Education Initiative) and CPVE (Certificate of Pre-Vocational Education) have directed their methodology to learning by doing, learning how to learn, and forging links with the world outside the classroom. But it was the GCSE which gave the force of examination requirements to a philosophy of active learning spread across the whole of secondary education, covering many subjects besides English.

The significance of the English educational revolution clearly affects teachers and teaching methods. Obviously, there are still many circumstances in which teachers should stand up and explain and teach, but the emphasis is now on partnership and a sharing of responsibility. The new examinations will not work unless the learners are active participants in the learning process. Thus, the emphasis is on processes. It also follows from this that teachers will be venturing into the role of observers of events they organize. And observation can merge into research. By this I do not mean that teachers will start producing academic papers and research reports, but that they will note what goes on, look for patterns of behaviour, and seek to modify their assessments and their teaching methods as a result.

In view of the huge disruption involved, it might have been expected that teachers would be the main opponents of the revolution. But this has not been the case. Although they have criticised the government funding for being inadequate, they have generally welcomed the changes in educational methods, embraced the opportunities for classroom assessments and endorsed the new emphasis on oral and group skills.

Although in British classrooms the revolution may be regarded as a purely local phenomenon, this is highly unlikely to be the case. A legacy of the British Empire remains in the educational systems of many countries in the world. This legacy has often been grudgingly accepted and frequently criticized, mainly on the grounds that it is too academic and does not take into account the need for practical experience which is so valuable to developing nations. Teachers have complained that by the time the children have worked their way through the textbooks there has been little time to experiment. True, the builders of the British Empire did not discriminate, and the deficiencies of the academic curriculum were just as evident in Leeds as in Lagos. But because the legacy lingers on, the English educational revolution is bound to be welcomed by many teachers and educationalists in the Commonwealth.

Nor are the shock waves likely to be confined to the Commonwealth. In the rest of the world the English educational revolution based on process rather than on product, on doing rather than analysing, will not go unnoticed. While it may not be copied in any great detail outside the Commonwealth, it is certain to be influential and may well set the tone of the debate on curriculum development.

It is within this revolution that simulations and similar techniques are due for re-evaluation. In Britain it is no longer possible for secondary school teachers to say, 'Yes, we think these techniques are valuable, but we can't use them this term because of the examinations.' The GCSE is not possible unless the techniques are used, because the techniques are now part and parcel of the classroom assessment. Moreover, no school would contemplate entering their pupils for the written papers or the externally examined oral tests if they have been receiving an exclusive diet of talk and chalk.

The conclusion, therefore, is that simulations, group exercises, role play, improvisations and so forth now have examination status. Their importance is immeasurably enhanced. The implications of their use is something which requires consideration, observation, and research. It raises many questions. How should simulations be chosen? Can teachers and/or students design their own, and if so what should be aimed for and avoided? How can oral communication skills be assessed in non-taught events? What is the best way to tackle the de-briefings after the events? Can exercises and role play be modified in such a way that they become simulations? How can an art lesson, or a history lesson, or a geography lesson incorporate a simulation?

The object of this book is to examine these questions in the light of experience. It is not unknown territory. Simulations and other non-taught events are part and parcel of the techniques which some teachers have been using for many years as part of a child-centred philosophy in which the learner is an active participant in the learning process.

Chapter 3
Design

Creativity

There is a story about a sculptor who had just finished making a stone elephant. He was asked, 'How did you do it?' He replied, 'I chipped away those bits that did not look like an elephant.'

Simulation design can be rather like this. Most people think of creative work as building, but it might be just as useful to think of it as chipping away. The simulation designer starts with a thousand potential simulations and ends up with one.

Creation involves closing options. Should the simulation be closed or open-ended? Should it be on a national or a local level? Should there be group roles or individual profiles? As each of these questions is decided, a great many possible simulations have been chipped away.

When a teacher inspects a package of simulation material, he or she may think that the author started off with a nice clear objective and the simulation was arrived at by some sort of logical deduction. What the teacher does not see is the author's wastepaper basket.

Authors must speak for themselves on this question, but I have not met one who regards creative writing as a simple matter of moving from a starting point, which is an objective, to a conclusion, which is the finished product, rather like a motorist plotting his journey on a map. Creative thought processes seem particularly difficult to recall afterwards — it is not like remembering what one bought at the supermarket. It involves a great deal of appraisal, discarding, selecting, and altering, and sometimes changing things around completely because of some new idea.

It is difficult to describe what is meant by design in a simulation. Some simulations have style, elegance, wit, and are not unnecessarily complex. A small number have a touch of genius, bringing a gasp of admiration for their ingenuity or profundity. A well-designed simulation has balance, the parts fit together well, and there is sometimes a degree of ambiguity, allowing different assessments of its meaning and impact.

With simulation writing, authorship is not usually enough; there has to be a testing of the product. Unlike writing a novel or a poem, the author cannot simply submit it to a publisher or ask for an expert opinion — it has to be tried out — and it is unusual for a simulation to be right first time. Often it needs considerable alteration and re-testing.

So what advice can be given to someone who wants to design their own simulations? A cursory glance at the academic literature on simulation design will reveal a wide variety of step-by-step recommendations, usually in the form of a flow diagram of sequential steps, something on the lines of:

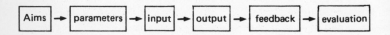

This seemingly plausible guide is about as useful to designers of simulations as it is to designers of clothes or ocean liners. It is not just that empty labels are strung together, but that they are sequential and therefore restricting. Why should aims be completed before input is considered? Although feedback can occur only after something has been produced, why should not the results affect the declared aims, or affect the parameters, input and output? In an effort to meet this objection some academics introduce a series of backward and forward pointing arrows, looking rather like the diagram of a railway network. However, a new objection emerges. As the arrows increase in number the diagram loses its flow and its sequence, and is little better than a vague checklist of individual tasks.

Even as a checklist it misses the point. Authorship is not about a series of tasks, it is about imagination, values, consistency and so forth. Some authors suggest that without a sequential flow diagram or a checklist of tasks there would be no guidance at all, leaving authors at the mercy of aberrant inspiration. This is a misconception. One can be systematic without being sequential. As mentioned in *Designing Your Own Simulations* (Jones 1984a) there are four basic questions:

1. What is the problem? — issue, situation
2. Who are are participants? — roles, identities, powers
3. What do they have to do? — job, decision-making, function
4. What do they do it with? — documents, materials, instructions, existing knowledge

These are not steps. Nor need an author ask and answer the questions explicitly — since all completed simulations contain implicit answers to these four questions — otherwise it is not a simulation. If the questions are asked as a conscious part of an author's methodology, then they can be asked in any order, at any time during the creative processes, and can be asked and answered more than once.

More important than these questions are the criteria of what makes a workable simulation — consistency, plausibility, and interaction.

- — Consistency: Does the simulation contain inherent contradictions, for example between the aims given in the organizer's notes and the instructions on role cards?
- — Plausibility: Given the main concept, whether fact, fiction or fantasy, does it contain implausibilities, as when a vital piece of information is missing from the role card of someone who would know this information.
- — Interaction: Although the topic may be highly interesting and relevant, is all the action left to a handful of people while other participants have part-time or passive roles?

Of course, these are not the only criteria. A dull and routine simulation can be consistent, plausible and interactive. But the three considerations have the merits of focusing attention on authorship as a whole, not on individual tasks.

Most teachers who design their own simulations do so with the aims of increasing the knowledge and understanding of their own particular subjects. With subject-orientated design it is easy to overlook the importance of consistency, plausibility and interaction. A common temptation is to write a lot of facts into the documents. This can result in inconsistency between doing the job and learning the facts. It can result in such implausibilities as having to fill in fact sheets as a check that facts are being learned. There is the danger that the simulation will be unbalanced, with the role cards giving work to a handful of key participants and the rest being passive or part-time.

Simulation design is neither easy nor difficult in itself. It depends on how it is done. It can be extremely easy, and children have no problems in designing their own simulations. They do it at an early age, an act of authorship which usually goes under the misleading title of 'play'. Their answer to the fourth question 'What do they do it with?' is usually existing knowledge. Their existing knowledge of hospitals helps them design simulations involving doctors and patients, television helps them with cops and robbers, and what mother does helps them in deciding how best to serve tea to the dolls.

One benefit for teachers (and pupils/students) who have a go at designing their own simulations is that it helps build up experience of what works and what doesn't, what is easy and what is difficult, what is interesting and what is dull. Such practical experience can be of considerable value in developing one's own critical faculties, which is of particular help when choosing from a selection of ready-made simulations.

31

Like everything else, appreciation of simulation design requires experience. The five simulations which follow serve to illustrate the diversity of design. Each has about it some special feature which, although not making it a typical example, provides an interesting case to study.

The last two examples, SPACE CRASH and THE LINGUAN PRIZE FOR LITERATURE illustrate two other aspects of simulation design:

(a) the advantages of a sequence of simulations, and
(b) a view of the author's wastepaper basket.

The first point is fairly self evident — if a teacher is designing one simulation then why not design two or three? The later simulations will give participants the chance of putting into practice the lessons learnt through experience of the first event. A similarity of format should make the authorship easier in the second simulation. It would reduce the amount of reading-in time for the participants, and help them to behave more effectively.

Less obvious is the wastepaper basket. Some simulations seem so simple and elegant that they appear to have been dashed off in half an hour after a stroke of inspiration. Authors rarely mention the wastepaper basket. In articles and books about simulations the attention is given to the final product and this reveals little or nothing of the processes, the changes, the rewritings. Authors' wastepaper baskets usually overflow, and contain maybe ten times more words, thoughts, and concepts than the final version.

TENEMENT

TENEMENT is one of the best known of British simulations and was written by some of the staff of the housing charity, Shelter. (The authors are not named in the simulation materials.) It concerns the problem of families living in a multi-occupied house in a large city. There are 14 roles: seven tenants, six agencies, and one landlord. Each has a role card of about 500 words. There are no other documents except for the organizer's notes and some chance cards which the organizer hands out if and when they might be useful.

It is intended for young people of 14 and upwards and the stated purpose is 'to make young people aware of the difficulties and frustrations of living in such a situation, and to point to ways in which some of these difficulties could be solved by the introduction of agencies concerned with such problems'.

Looked at from the point of view of design, TENEMENT is particularly interesting as an example of excellent and emotive subject matter linked with a design which is weak on participation. The main problem is that although the tenants can visit whichever agency they think might be most useful in their individual circumstances, the participants who are in charge of the agencies might be overworked or have nothing at all to do. The participants in charge of the Citizens' Advice Bureau may sit at their table throughout the simulation twiddling their thumbs if the tenants decide to home in on those agencies which seem to have more relevance. The Department of Employment participants are also likely to have a long wait. They have only one potential applicant, Mr Johnston, as he is the only unemployed person. However, Mr Johnston and his family are the only tenants facing immediate eviction, so he may be too busy seeking help from the Rent Tribunal or the Housing Department ever to get round to visiting the Department of Employment.

The organizer's notes are of little help since they do not mention the participation problem. They contain about 1500 words — half of which consist of suggested questions for the de-briefing.

Suppose that two or three of the agencies are sitting around feeling restless with discontent or bored with inaction. Should the organizer refrain from interfering and hope that things sort themselves out, or walk over to the agencies and commiserate with the staff on their idleness? Or should the organizer search through the chance cards and, finding 'You have been made redundant', give it to the participant at the tail of the queue at the Housing Department?

Some of the chance cards might themselves cause bottlenecks. There is one for the landlord saying that a property developer has offered to buy the house, but only if there are no tenants. And if the landlord decides to try to evict the other six tenants, they will probably join Mr Johnston in a queue, or else all the tenants will start to negotiate with the landlord personally, which would mean that all the agencies would be without customers.

Nor is poor participation the only design problem. What are the simulation facts? The organizer's notes say nothing about this. Yet the Rent Tribunal's role card says, 'Remember, it is important to get all the facts. The tenant and the landlord may tell different stories, and the Tribunal will have to decide which one is nearer the truth.' But there are no authoritative documents or facts for the Rent Tribunal as the role cards are the only source of information. But lies are anticipated and the Williams family, for example, are instructed on their role card to conceal the truth about being evicted from their last council home and about the fact that they still owe rent. The door is open for lying and cheating, allegations and counter-allegations, and what may start as a simulation may well end up as an informal drama or a general free-for-all.

One of Shelter's education officers said that on one occasion the tenants tried to murder their landlord and the Shelter official had to be called in to act as policeman to try to control them. On another occasion the old-age-pensioner got down on his knees and proposed to the unmarried mother in order to share her flat, and one tenant decided to emigrate to Australia (Jones 1974).

It is not recorded what the participants at the agencies were doing while the tenants were having fun and games, but it seems clear that some of the participants were turning their imaginations to good account.

This demonstrates not only the disadvantages of a poorly designed simulation, but also the strength of a simulation which evokes sympathy and concern and which the participants may decide to rescue by transforming it into an informal drama. Unfortunately, not many simulations are as evocative about injustice and poverty as TENEMENT. If they are poorly designed, then any rescue attempt may be in the form of playing it for laughs.

STARPOWER

Garry Shirts is an American author of several highly ingenious and evocative simulations, including STARPOWER which is the world's most used simulation. It too is about poverty and power, but in both content and design it is very different from TENEMENT.

STARPOWER has what is known as a hidden agenda. Things are not what they seem to the participants. Ostensibly it is a sort of trading game. The rules say that the primary goal is to become wealthy but that the participants can decide how to do this. Usually they decide in favour of competition. They are divided into three groups — squares, circles and triangles — and each participant takes his initial wealth from a grab bag. What is not realized by the participants is that most of the squares start with greater wealth than the circles, and that most of the circles start with greater wealth than the triangles.

After a few trading sessions the organizer intervenes to identify the groups according to their wealth. To be a square requires a specific amount of wealth, so a small number of people who were originally squares are demoted to circles, and one or two circles are promoted to the top group, the squares. Similarly, a few promotions and demotions occur among the circles and the triangles. It looks as though the ups and downs were the result of trading skill, whereas in fact it was inevitable from the start because of the handout of original wealth.

Already the behaviour of the participants has changed. Each of the three groups develops a solidarity and places a distance between itself and the other groups. The squares tend to walk tall and smile, whereas the triangles tend to slouch. Even the organizer seems to prefer the company of the squares to that of the other groups.

More trading sessions occur and then the organizer says that, as the squares have been doing so well in their trading, they are to be rewarded for their skills by being allowed to change the trading rules if they wish. This is a touch of genius by Garry Shirts. The idea that those who do well in the game can then change the rules half-way through is mind-boggling.

What happens next is unpredictable. Usually the squares consult with each other in their group and decide to alter the rules in their favour with the aim of preserving or increasing their wealth. In the meantime, the circles and triangles are discussing the extraordinary turn of events in their groups and are becoming more and more indignant.

After this it rarely takes more than one or two brief trading sessions for the dissidents to mount a campaign of protest. At the right moment the organizer steps in and, pointing to an angry, gesticulating member of the under-privileged groups, cries 'There is your revolution!' Or words to that effect. End of simulation. Everyone now realizes there was a hidden agenda. Many realize that the emotions and attitudes they have been feeling have been contrary to their normally stated beliefs. The squares realize that they have not only been using power, but also enjoying it. The triangles realize what it is like to be on the wrong end of injustice and inequality. The circles, who may have supported the class structure in the hope of bettering themselves and thinking that at least they were better off than the triangles, may be feeling rather guilty. And guilt often brings counter-attack — 'Is STARPOWER supposed to be an attack on our competitive society, on capitalism, on our democratic way of life?' etc. (It would be interesting to see STARPOWER used in a communist country.) Nor are the personal bitter episodes soon forgotten, even though they may be forgiven. Even if, at the crucial moment, the squares become altruistic and decide to alter the rules in favour of more equality of wealth, this in itself can result in a bitter and hostile reaction from the other two groups. 'Why should you be able to give us things — clear off!'

Sometimes the participants learn that violence is not the prerogative of children or non-intellectuals, and simulation literature contains some extraordinary descriptions given by the organizers or observers at STARPOWER sessions. Coleman (1977) gives an account of what happened at one session at a teacher training college in Cambridge:

> On one occasion a group of leftish liberal studies lecturers announced 'The name of the game is GRAB', and very shortly afterwards I was knocked to the floor and a pack of bonus cards torn from my hand. This was a pity — a meeting intended to show the hidden violence of our established society showed intead only the boorishness of some of its opponents.

STARPOWER is simple to introduce, highly ingenious, a remarkable opportunity for self-revelation, and as explosive as gelignite. It needs careful handling. The organizer must be able to deal with the intense inter-personal hostilities which may well occur. As Garry Shirts says, STARPOWER should probably be used only by 'teachers who feel comfortable with vigorous reaction'.

DART AVIATION LTD

DART AVIATION LTD is another good example of a fully participatory simulation, and although very different from STARPOWER it shares the qualities of simplicity, ingenuity, and some tongue-in-the-cheek wit. Like STARPOWER, it was designed by an experienced author, Charles Townsend, and, like TENEMENT , it is a British simulation. It is the third in a series by Townsend entitled *Five Simple Business Games* — the other four being GORGEOUS GATEAUX LTD, FRESH OVEN PIES LTD, THE ISLAND GAME and THE REPUBLIC GAME. But although simple, it is not short and lasts about four hours involving roughly seven rounds of manufacture and trading.

It is intended for non-specialist pupils in the fifth and sixth forms of secondary schools. This, together with its simplicity, makes it rather unusual since most business simulations are somewhat complicated affairs, requiring several hours of pre-reading and designed for adults at management level.

The title is slightly misleading, as there is not just one company called DART AVIATION LTD but four, all starting with identical resources, operating under identical trading conditions and competing against each other.

Also very unusual is the combination of a business decision-making simulation with the manufacture of an actual product. The participants manufacture paper darts. This is a nice touch of humour, since paper dart making is the classic symptom of boredom. The darts must comply with specific quality control or they are rejected. They must be a specific length and width and fly at least four metres. Raw materials and tools are available at a fixed price. There are two sizes of paper and adhesive tape is purchased by the metre. Scissors and rulers (the tools) may be hired at £1000 each for each manufacturing period they are required. Labour is also an expenditure. The organizer uses a stop-watch or a watch with a second hand and the labour costs are £10 per worker per second.

DART AVIATION LTD has a nice balance between manufacturing and selling. Sales of aircraft depend on orders received, and orders received depend on what price each company puts on their aircraft and how much they spend on advertising. As in most other business simulations involving market decisions, the organizer has arithmetical tables to work out the answers. Thus, if one company spends the maximum amount allowed on advertising and prices the aircraft at the minimum amount the rules allow, it will receive more orders than a rival company which spends less money on advertising and sets a higher price per plane.

If, however, a company receives more orders than it can fulfil, these surplus orders cannot be held over to the next trading period — they are lost. So a firm can sell no more than the number of planes it manufactured. If a firm manufactures a surplus of planes, these can be held over in stock until the next round.

From this brief outline — there is a lot more to it than this — it can be seen how a simple idea can generate a great many tasks which are common to most businesses: accounting, deciding how much raw material to buy, deciding on advertising and price, productivity, product design, training, etc.

Depending on the size of each team, it is recommended that individuals should allocate specific jobs among themselves — managing director, sales and marketing manager, accountant, worker, and so on. Since the motivation is high — the companies want to succeed — this division of labour within teams can enhance the participation. A bright participant who finishes a task first has a strong incentive to help others, and is unlikely to sit back and look bored. Although one role is 'worker', that does not mean that only one participant in a company can manufacture the aircraft. If two or more workers seem the best solution, then this can be done.

Similarly, if one team finishes its decision-making well ahead of the others in each round, it is unlikely to stop dead, since there are still problems to solve, methods of accounting to be explored, and future policy to be discussed.

Like STARPOWER, but unlike TENEMENT, DART AVIATION LTD has balance and fits neatly together with all the parts having both opportunity and incentive to work. The danger with DART AVIATION LTD, like other business simulations which have a series of trading sessions, is that the organizer may let it run on for too long. There usually comes a time in such a simulation when the teams have solved their problems and worked out their best strategies, and things tend to become routine and repetitive. But this is a question for the organizer to decide; it cannot be laid down in advance as the abilities and interests of different groups can vary considerably.

SPACE CRASH

I designed SPACE CRASH in the same format as my earlier RED DESERT and SHIPWRECKED, which were the two parts of SURVIVAL, the first of my *Nine Graded Simulations* (Jones, 1984).

37

The design similarity meant that there was no reason for me to invent the basic idea, which was that of participants being in unexplored territory with no map. The map is built up square by square according to the direction of travel. Each map square has written round the four edges what can be seen in the four directions — grass, flatland, etc. — and each direction has the number of the adjacent square. Each role card contains information which the others do not have. In order to reach safety, the group has to co-operate and exchange information and ideas. If a group ignores or squashes one of its members, then this is not only impolite but is inefficient.

SPACE CRASH is the first of *Six Simulations* (Jones 1987b). The aim is to give opportunities for language and communications skills. The six are published in one book with spiral binding to help in photocopying the documents for classroom use.

SPACE CRASH was selected as the first of the series because the event is extremely easy to organize; it does not take very long, the amount of information for each participant is small, the procedures are straightforward and self-evident, and there is usually a very high level of participant interest. However, survival is extremely difficult. Almost all groups die. This is not a matter of luck; it is because groups do not use sufficient common sense and do not communicate effectively. Failure to survive convinces the dead space crew that they were on their own and that the organizer gave them no help. It is a useful lesson to learn, not only from the point of view of the participants — it can also be a novel experience for the organizer, and serve as a basis for non-intervention in the subsequent simulations.

The original version of the notes for participants (which was reproduced in the first edition of this book in 1980) and the 1987 version, are shown on pages 40 and 41, respectively.

In the 1987 version I made a few stylistic changes for reasons of clarity. But the main reason for rewriting the document was the addition of two pieces of practical advice, plus one major conceptual change.

One of the practical points is the instruction to lay out each new map square in position according to the direction of travel. I had not included this in the original version because it seemed so obvious. Although most groups laid out the squares correctly, there were a few occasions when teenagers had laid out the second and third squares in random locations, rather as if they were laying out cards for fortune telling.

Another point of practical value was the requirement for scrap paper (also added to the organizer's notes). This was introduced because experience of running SPACE CRASH had revealed that scrap paper could be useful for (some) members of the space crew, helping them to think, plan and take notes, depending on their own personal work habits. There seemed no reason why space crew should not have scrap paper.

The major change concerned the main problem with SPACE CRASH. Because it has cards (map squares) it has a superficial resemblance to a game. In the original version a warning is given in the notes for participants:

> SPACE CRASH may look like a game of chance but it is not. If you treat it like a game, you are not likely to live very long.

In most cases this warning was sufficient. Participants did behave professionally. But occasionally, perhaps because of the terminology of the organizer, participants kept referring to it as a game in the de-briefing, and they evidently thought of it as a game during the action, which did them no good at all for the following reasons:

(a) it led them to concentrate their attention on the map squares, the seemingly gaming element, rather than on the non-gaming elements — the highly important role cards and Erid's diary;

(b) it diverted them from professional behaviour, from the need to explain and convey information, and from making sure that all the relevant data was exchanged and options discussed;

(c) since games are frequently associated with chance, this led them to believe that survival was a matter of luck, and this in turn led them to be careless. In the de-briefings they blamed their deaths on bad luck;

(d) since games are played for enjoyment the participants (or rather the players) tended to take unprofessional attitudes, joking and taking quick decisions to see if they could 'win', and they were overly impatient to find out what the next map square revealed.

One problem with the original wording of the notes for participants was that it put the idea of a game into their heads without explaining why it was not a game of chance, or indeed, why it was not a game of skill. The phrase 'If you treat it like a game' probably carried the meaning 'If you treat it flippantly'. However, many players do not treat their games flippantly, so the issue was wider than flippancy.

A characteristic of most board games and card games is that the player is detached and occupies a position above the board or the table, looking down, moving proxy figures or cards across the surface, like a puppet master or a god. I wanted to encourage the notion that there were no proxy figures, and that they were themselves inside the situation, not above it. I still stated that it was not a game but left out the phrase 'game of chance' and added the words 'playing' and 'players'. I also introduced the positive notion of professionalism in relation to decision-making.

NOTES FOR PARTICIPANTS

SPACE CRASH (1980)

What it's about

You are survivors of a space crash and it is your job to stay alive.

Five profiles — Andro, Betelg, Cassi, Draco and Erid — explain the situation. Read them carefully. They are important.

You have a map square made by Betelg which shows the area in which your spaceship has crashed. Around the edges of the map square is a description of what you can see one day's walk away.

When you have decided which way to go, tell the controller the number of the map square you want. Each new map square shows you where you are and what you can see from that square.

You have a diary in which to keep a record of your day-to-day movements from square to square. There are no diagonal movements and you must all stick together.

Advice

SPACE CRASH may look like a game of chance but it is not. If you treat it like a game, you are not likely to live very long.

Before you start deciding which way to go, try to make sure that everyone has told you everything they know about the planet.

Don't waste time trying to think of ways to 'break the rules'. For example, you have nothing to carry water in — and that's that.

It is a good idea to have a small coin or marker which you can move from square to square as you travel across the planet. It helps to show you exactly where you are and it helps you to keep an accurate diary of your movements.

SPACE CRASH

Notes for Participants (1987)

What it's about

You are space crew and your spaceship has crashed on the planet Dy. You are the only survivors and it is your job to stay alive. There are five role cards — for Space Officers Andro, Betelg, Cassi, Draco and Erid.

Erid will receive a twenty-day diary to record your day-to-day movements across the hostile planet. Betelg will receive Map Square number 1 showing the area of the crash.

You will also have pen or pencil and some scrap paper which may be of use when discussing which way to go.

When you decide which way to go, tell the Organiser the number of the map square you want. When you receive a new square, lay it out in its correct position next to the other squares, showing the route you are taking.

What you can do and what you cannot do

The only facts are in the role cards and on the map squares, and you are not allowed to invent 'facts' in order to survive. For example, you have nothing to carry water in and you cannot invent a water carrying container.

It is a good idea to have a small coin or marker which you can move from square to square. This will show you exactly which square you are on and it will help Erid to keep an accurate diary of your movements, particularly after the first few days.

Apart from the need to stick together and not to travel diagonally at any time, you can do anything you have not been told not to do, providing you accept your responsibilities as Space Officers.

Try to see yourself inside a map square looking outwards, not above a map square looking down. You will not be playing a game, you will be inside a dangerous situation. You are not players, you are professional space crew who search for options, exchange ideas, and exercise caution.

> Try to see yourself inside a map square looking outwards, not above a map square looking down. You will not be playing a game, you will be inside a dangerous situation. You are not players, you are professional space crew who search for options, exchange ideas, and exercise caution.

This point is reinforced in some of the role cards. For example, the last paragraph in Andro's role card originally said:

> I will tell the others what I know. And they must tell everything they know about Dy, or we will surely die.

The 1987 role card replaced the above with:

> We must remember that we are professional space crew and this is not a training exercise back in Space School. This is not a game or a puzzle. This is Dy the planet of death. We are not trying to win something, we are trying to avoid death. We must think, and talk, and exchange ideas.

As can be seen, the new version also states that the event is not a puzzle (puzzles have solvers, not participants) nor an exercise (exercises have no roles, otherwise they are simulations). It also contrasts the dangers on Dy with exercises in Space School, which not only helps to distinguish between simulations and exercises, but emphasises that failure means death, not just bad marks, It would have been possible to design SPACE CRASH as a simulation within a simulation, the location being Galactic Space School, but this would have reduced the awareness of danger, and might have led to an 'It's only an exercise' attitude, and subsequent carelessness.

In practice, almost all participants in SPACE CRASH find it exciting, involving and professional. They do not, of course, actually believe they are on Dy. In fact, the first map square is often regarded as a bit of paper, but after it has been joined by the second and third map squares, and after options have been discussed and ideas and information exchanged, the participants really do feel themselves to be inside the event, not above it.

SPACE CRASH, both in the form of documents and as an event, demonstrates the crucial distinctions between being a player and being a participant. It highlights the importance of trying out simulations in order to see whether an author's expectations occur in practice. The comparison between the two versions reveals the wastepaper basket.

THE LINGUAN PRIZE FOR LITERATURE

This is the last of the *Six Simulations* and has more diverse features than any of the others.

It may be useful to view THE LINGUAN PRIZE FOR LITERATURE from two aspects — as part of a series, and as an example of rewriting.

Although this simulation (or any of the others) can be used as a one-off simulation by teachers of English (media studies, life skills, business

THE LINGUAN PRIZE FOR LITERATURE

Notes for Participants

What it's about

The action occurs on the day of the awarding of the Linguan Prize for Literature. You will be an author, or an editor, or a judge, or a television journalist. There will be more authors than editors.

There will be a short television programme, *Arts at Noon*, which has a 12.00 deadline in which journalists can interview judges, editors, or even authors if they are not too busy writing their entries for the competition.

The awarding of the Prize will be televised in the evening from the banqueting hall of the Linguan Towers Hotel, Lingua City.

All this is explained in the documents. Authors and editors have a memo from the Linguan Publishers/Authors Committee. Judges have a memo from the Linguan Arts Committee. Television journalists have a memo from the Linguan Television Liaison Committee.

Everyone will have a copy of the announcement about the subject of the 10th annual prize, plus a copy of the page of readers' letters in the *Linguan Times*.

What you can do and what you cannot do

You can visit other groups and other individuals to discuss arrangements for the broadcasts and the interviews. Editors can move around according to which of their authors needs most help. Editors can assume that if they are interviewed on *Arts at Noon* then this would help promote the sales of their authors' works, and their bosses would be pleased. Judges can assume that if they themselves are interviewed then the publicity is likely to benefit their careers.

If you are being interviewed and are asked a factual question and do not know the answer, ('What was the title of your last book?'), then it is permissible to invent a plausible reply. But you cannot invent facts to increase your own prestige, ('I have been awarded the prize for the last three years.' 'I am the most famous author in Lingua').

The deadline for the broadcasts cannot be extended if the journalists are not ready. If they arrive in the studio only half a minute late they are likely to find that a substitute programme is already being transmitted and that their boss wants to see them immediately. The *Arts at Noon* programme must finish on time, but the television coverage of the prize-giving ceremony will be allowed to overrun.

studies), the rationale of the series is the building of one skill on others.
The subtitle to *Six Simulations* is 'An oral skills resource'. Briefly the
idea is as follows:

SPACE CRASH provides experience of working informally in groups
with no assistance;

MASS MEETING gives experience of public speaking in the context of
a trade union meeting about a letter from management offering to
buy the members out of their union;

THE RAG TRADE involves both creative design and judging of a
fashion theme for the spring;

BANK FRAUD involves office politics, ethics, and confidential
discussions in pairs. Two executive managers have the job of
appointing a team to investigate fraud.

TELEVISION CORRESPONDENT includes ministers who give a news
conference and are interviewed afterwards in the news spots of
various programmes.

Most of these skills resurface in THE LINGUAN PRIZE FOR LITERATURE —
television journalism, group work, public speaking, organizing. However,
the event is more creative and diverse than the other five simulations,
and it deals with literature as well as journalism.

In all six simulations there are notes for participants. All have realistic
documents: memos, letters, extracts from newspapers. Apart from the
notes for participants, which are not part of the action itself, the only
non-realistic documents are the role cards in SPACE CRASH, and even
here the information is written from the inside, 'My name is . . .' rather
than 'Your name is . . .' One of the advantages of having realistic
documents is that they are flourishable — 'This memo says . . .', 'This
letter makes the point that . . .'

It can be seen that with *Six Simulations* the skills are developmental,
and the situations are diverse. These characteristics can be applied to
any group of simulations, or the simulations can be variations of the
same subject theme, as in Townsend's *Five Simple Business Games*.

Superficially THE LINGUAN PRIZE FOR LITERATURE may look
complicated from the outside because there are four distinct functional
roles — author, editor, judge, journalist. However, it is simple when seen
from the inside since each person has only one job. As explained in the
organizer's notes;

The skills relate to literature, emotions, and human values. The authors
create their works, the editors enhance and publicise them, the judges
compare them and the television journalists present them. All four roles
offer scope for imagination and understanding.

Originally there were two working documents. One was a page of

Publishers/Authors Committee

MEMO

From: Publisher/Authors Committee

To: Authors and editors involved in the prize for literature

All publishers have agreed to our suggestion to allow their
editors to be interviewed by the media, so as to help the public
appreciate that publishing depends on authors who submit their
works to publishers, and who are then helped by the advice
and suggestions of editors.

A survey we have just carried out reveals that 91.3 per cent
of those questioned had only the vaguest idea of what a
publisher does. A total of 61.9 per cent thought that a
publisher was the same as a printer. Misconceptions
abounded. One person said "A publisher publishes books and
also publishes titles." Another said "A print run is a
run of ink which messes up the pages". Only 15.5 per cent
realised that a print run had something to do with the
number of copies of a particular book ("a title") run off
by the printing presses at one particular time.

Please feel free to disclose the results of this survey if
you are interviewed, and please take the opportunity to talk
about the creative aspects of authorship and publishing,
and about the close links between an author and a publisher's
editor to help the work along the path to publication.

letters in the *Linguan Times* and the other was an announcement by the Linguan Arts Committee that the subject for this year was 'Love'. Also, the original version of the notes for participants referred to 'publishers' rather than 'editors'.

The simulation was tried out with 11-year-olds and upwards, and these trial runs revealed, among other things, that the word 'publisher' caused misunderstandings. Participants, even graduates, were far from clear what publishers did, and tended to assume that one of their jobs in the simulation was to publish something. Of course, they did the job of editing, advising, and were interviewed on television, but they had a sense of unfulfilment.

Another problem occurred when I tried the simulation with adults at an annual conference of the Society for the Advancement of Games and Simulations in Education and Training. In order to avoid embarrassment to professional people I decided to pair authors and publishers, having the same number of each, so that any one could say to their partner, 'I'd prefer not to be author, may I be publisher and you take on the job of writing something?' This proved to be a mistake. There was no embarrassment. Some publishers felt that they were rather redundant, and in one case a pair decided to constitute themselves into two authors rather than author and publisher. A full description of this particular event is given in Jones (1985b).

The difficulty was dealt with in three ways. The title 'publisher' was changed to 'editor'. It was stipulated that there would be more authors than editors. Memos were introduced to spell out the functions, to provide additional facts, and to give advice. However, there were three memos, not four. Each author and editor received a joint memo from the Publishers/Authors Committee in order to emphasise close personal co-operation.

Comparisons of design

Most simulations are non-practical and two dimensional in the sense that they produce nothing other than decisions written on paper. They are usually talking shops. Hands are used for holding pens and for voting. There is nothing wrong with this format, and many good simulations are forums of discussion. However, the five simulations singled out are not formal committees debating an issue, followed by a decision. All, in their own ways, are emotional personal experiences, involving attitudes and ethics, and all produce something other than paper.

In TENEMENT there is a high degree of personal involvement and emotion. Decisions are reached, but the issue is personal poverty, not, for example, an abstract review of what might be done to relieve hunger in the third world.

STARPOWER is one of the most emotional simulations ever devised. Like all good simulations, the participants are not instructed to be emotional; the emotions arise out of a class system of power, and out of the sudden granting of additional power to a favoured group. There have been occasions in both STARPOWER and TENEMENT when physical violence has broken out in the classroom.

DART AVIATION LTD may give the impression of being something on the lines of a kiddies' handicraft lesson, but this is not the way the participants view it. The task of designing and producing darts which, when tested, crash to the ground or soar successfully can be a bitter or sweet emotional experience.

SPACE CRASH has also to be experienced to appreciate the emotional involvement. The personal involvement, the doubts, frustrations, arguments, anxieties and almost inevitable death cannot be appreciated by merely inspecting the map squares and role cards.

THE LINGUAN PRIZE FOR LITERATURE produces the literature of emotion, the anxiety of producing live television, the involvement of judges and editors. The feelings are markedly different from that of a committee simulation.

Comparing developmental aspects, the criterion for *Five Simple Business Games* is complexity of information. In a way DART AVIATION LTD is the odd man out in the series because the others are essentially committee simulations based on several rounds of trading. No person to person trading takes place. It is all theoretical. What happens is that decisions are made about figures for production, price etc. and these figures are then compared by the organizer and an arithmetical calculation is made about how big a slice of the market is won by each company, what the profits are, and so forth. The number of business factors which have to be compared increases during the series, but the basic method of arriving at a 'solution' remains the same. Thus DART AVIATION LTD can best be regarded not as part of a series, but as an example of a production line simulation. These have become quite popular, and groups form into teams, each planning, manufacturing and supervising an actual product — perhaps a cardboard container, or a construction of interlocking bricks.

Although not part of a packaged series, STARPOWER can well be regarded as part of the Garry Shirts series, the criteria being surprise, ingenuity, and personal involvement. In STARPOWER actual trading takes place, as it does in Shirt's BAFA BAFA (1977). In his simulations Garry Shirts deliberately distorts reality to make his points (Shirts 1975). In some cases it makes more sense to categorize simulations, like novels, according to their authors rather than according to their subjects. For example, the subject matter of SPACE CRASH and THE LINGUAN PRIZE FOR LITERATURE are miles apart, but both are similar in style, and also both aim at developing language and communication skills.

The simulations described in this chapter are all relatively easy to obtain. They were frequently tested before publication, which cannot be said of all published simulations. Some simulations contain so many design faults that one wonders if they were ever tested, or maybe they were tested once, then changes were made, and the final version went untested in the expectation that in practice things would work out as planned.

As far as classroom-designed simulations are concerned there is no need to test simulations, just run them. Of course, it may be useful to modify the version if a rerun is desired, or if the teacher wants to use it again with another class. But basically, the testing possibilities in schools and colleges are extremely limited. In any case, homemade simulations are not expected to be highly professional, and part of the enjoyment is seeing what happens. If, for example, one group is given only part-time roles, then the criticism after the event, or even during it, is likely to be much more lively than if the participants are criticizing a simulation taken from a shelf. It is more interesting and educational when someone says 'Your group did not design that very well' rather than 'That published simulation is not very well designed.'

It is easy in a class to set up design teams, and subsequently the designers can organize and run their own events. If time prevents more than one such simulation being run, then a vote can be taken. If the answer to 'What do they do it with?' is existing knowledge, then most children can design a workable simulation in five minutes, on the lines of:

> One group are managers of football clubs and the other group are the police, and they meet to discuss soccer hooliganism.

> A news conference is given by X, Y and Z because they are the experts on computers (goldfish, ballet dancing, tree climbing) and the others are reporters from local newspapers who want to find out what the local community is doing.

> There is a television debate between those who want some waste land to be made into a pleasure park, those who want it to become a factory site, and those want to turn it into a bird and wild life sanctuary.

Other examples of the design of instant simulations are given in the chapter 'Getting Started' in *Designing Your Own Simulations* (Jones 1985a).

Chapter 4
Choosing simulations

Problems

One of the main problems in choosing a simulation is that neither the reading of the organizer's notes nor an inspection of the documents can reveal the flow of interaction which occurs during the event. Viewing a simulation from the outside is not like reading a poem. The best part of the simulation – the action – is missing. Some of the most interesting and imaginative simulations contain all sorts of clues, suggestions and options which are deliberately buried at various levels, and these finer points of design are rarely mentioned in the organizer's notes.

If it were merely a case of difficulty of evaluation it would not be too bad. One could then accept the lack of data, and be open minded. Unfortunately, teachers who have little if any experience of simulations almost inevitably jump to a conclusion which is unfavourable. Looking at the materials from the outside puts them off. If there are a lot of documents, despite the fact that most of them may be read only by one or two participants, the simulation is judged as being too complicated. If, on the other hand, there are only one or two documents then the event looks empty and the reaction is that the participants would not know what they were supposed to do.

Even teachers who are experienced in running simulations often have difficulty in guessing how well a particular simulation would work in practice. There is no real substitute for participation. However, if the organizer's notes inform the teacher that the event requires a couple of hours to run, then it is unlikely that a full participatory trial would be possible. The usual advice is that a teacher should read the documents and then take on a role and imagine the situation as seen by that person. Although this is certainly better than a non-imaginative approach, it still means that the first step is to read all the documents, and this can interfere with the imaginative participation. Another way of tackling the problem is described in the organizer's notes to THE LINGUAN PRIZE FOR LITERATURE.

> Read the Notes for Participants and step inside the event by taking a role and reading only the documents which that person would have. For

example, take on the role of an editor. Read only the memo from the Linguan Publishers/Authors Committee, the announcement entitled 'The 10th Annual Linguan Prize for Literature', and the page of letters in the Linguan Times.

You learned (at the briefing) which of your authors are submitting entries for the prize, and you walk over to one of them who is sitting with pen in hand gazing at a blank sheet of paper. 'Hello' says the author, 'You are my editor, right?' To which you reply 'Yes . . . . .' and what else? Don't conduct your imaginary conversations in indirect speech; use the actual words that the author would hear. Now switch to the role of author. You heard what your editor said. How do you respond?

Take another role, say that of judge. You sit with the other judges who have just read the memo from the Arts Committee. One of them turns to you and says 'What should we look for this year?' And you say . . . ?

After a few such imaginary conversations try to find a couple of friends or colleagues who can spare 10 minutes and you can give them roles and ask them to converse. This experience will help you run the event confidently.

Facilities checklist

Before shopping for a simulation it is useful and time-saving to make a mental inventory of the conditions and facilities which will be available. The greater the facilities, the greater will be the area of suitable choice.

MONEY

Like shopping for anything else, the teacher is looking for value for money. The educational value of a simulation is not always reflected in the price, or the weight of the documents, or in the length of the list of worthy aims, or the gloss on the package. As with any sort of shopping, the more experienced the buyer the greater the chance of a bargain.

An important financial consideration is whether the publisher has given permission for documents to be photocopied for classroom use. At one time permission was hardly ever given, but in recent years several publishers have made this concession. Permission is usually restricted to the school or college, and usually includes only the documents for participants, not the organizer's notes.

The books or packages of simulations should make it clear whether photocopying is allowed, and if not whether the materials are durable or whether they will be used up by record keeping and form filling. If the simulations allow photocopying and are in book form, then it is useful to note whether or not the book has a spiral binding. A stiff spine can make photocopying difficult, and there is the possibility that repeated photocopying may cause the pages to come adrift from the spine.

NUMBERS

Most simulations are extremely flexible in the number of participants that can be accommodated. However, it is useful to check whether the simulation can deal with the number of students in the class. This information should be featured prominently, and should include minimum and maximum figures.

Some organizer's notes contain options for running the event with numbers that are larger or smaller than the ideal number, but if these options are not spelt out, then the following considerations may help.

In the case of a larger number of participants than roles there is a choice between (a) participants sharing roles, and (b) running the same event simultaneously in different groups.

Sharing roles is usually no problem, but in some cases there is a danger that the resulting group will be too large. For example, a committee simulation with 10 roles could be an unwieldy committee if increased to 20 members, and unless the time allowed for the meeting was also doubled then there would be less time for individual participants to make their contributions.

Simultaneous simulations are easy to arrange — say one group in each corner of the room — but there can be problems of noise interference if the event concerns public speaking rather than informal discussion. On some occasions it is possible to link simulations which are run simultaneously. For example, there could be several appointments boards, but instead of each having their own separate candidates there could be a group of candidates who could individually rotate from one interview to another.

With fewer participants than roles the options are (a) to abandon some roles, and (b) to have one or more participants taking on two roles. Which option is preferable depends on the simulation and the type of roles. For example:

TENEMENT. The tenants have individual names. It is more plausible to eliminate one tenant than to have one participant becoming two tenants. In the case of the agencies the question is more open. In some circumstances it may be better to abandon an agency rather than have one person run two agencies, but if one participant is competent to cope with two agencies, and if the length of the queue at this double agency is not a major consideration, then sharing an agency might be the best course of action.

STARPOWER . This will not work with fewer than three groups, so small numbers means a reduction of participants in each group.

DART AVIATION LTD. A small number of participants means a choice between having only one production company (and no competition) or having two competing companies, but each having only a small number of employees.

SPACE CRASH. Each of the five role cards contains important information. This means that with only four participants a role card cannot be left abandoned on the organizer's table. One of the participants must be given two cards.

THE LINGUAN PRIZE FOR LITERATURE. The minimum number required to make the simulation work effectively and interactively is eight (three authors, one editor, two judges, two journalists).

TIME

The materials should give an idea of the average amount of time needed and whether this includes the briefing and de-briefing.

As with numbers, time is fairly flexible in most simulations. If the time required appears to be too long in an otherwise suitable simulation, it is often possible to reduce the briefing time by a different form of presentation, or reduce the overall time by greater efficiency in the general running of the simulation. Similarly, a short simulation can be lengthened by various means, including a more thorough preparation, and an elaboration of some of the formalities and procedures within the simulation.

It may be a mistake to assume that the less able pupils need more time than brighter pupils to deal with a simulation. They may take more time to read and understand what they are supposed to do, but in the action part the brighter pupils spot more opportunities and will delve, prod, argue and discuss.

Time has a value, but there is a danger, as with money, in assuming that the less spent the better. Some authors argue the opposite. Elgood (1976) writes:

> Time is not at all a bad measure of value, and if certain items of knowledge are thought to be worth a major allocation of time, then the message of their importance tends to get over more strongly... It is too seldom admitted there is a positive correlation between the thoroughness with which people learn a thing and the effort they expend in doing it. Intellect is only one of the characteristics that is involved in the learning process. There must also be an emotional evaluation of the importance of the knowledge and this tends to relate to the emphasis placed upon it by the circumstances in which it is offered.

ABILITY

The main difficulty in assessing whether the materials are on the right level for the students' abilities is that these abilities cannot be assessed with any certainty without the students taking part in several simulations first. So if the teacher knows the students' abilities in simulations, all well and good. But if the teacher's views on student abilities are based on what happens in teacher-student orientated learning, the assessment may be inaccurate. Teachers who have had little or no experience of using simulations tend to underestimate the ability of their students to cope with the materials — not so much in the first simulation where there may be teething troubles, but in the subsequent ones. Even during the course of the first simulation it is often possible for the organizer to notice the gradual (and sometimes not so gradual) increase in skill and confidence of the participants in handling the situation in which they find themselves.

The problem becomes somewhat easier if the simulations are graded. This technique is becoming increasingly popular, with simulations linked together like steps, the early simulations being relatively easy and the others progressively more difficult. Sometimes simulations are graded within themselves, with preliminary exercises, trial runs, or simplified versions of the main simulation activity.

Another consideration is the possibility that individual students who have low ability or a limited knowledge of the language might be thrust into an embarrassingly prominent position. Role allocation is dealt with in the next chapter, but generally speaking simulations involve group work and work in pairs. This provides opportunities for shelter.

In the informal but hazardous atmosphere of SPACE CRASH there is nothing to stop participants showing each other their role cards (or to stop a keen individual grabbing the role card of another member of the space crew). In TENEMENT the agency workers and tenants can reveal documents. In THE LINGUAN PRIZE FOR LITERATURE the judges and television journalists can help each other, as can the authors and editors.

Unlike the normal classroom situation, individuals in simulations are less at risk of personal exposure. To be asked a question by another participant is less fraught than if the questioner is the teacher. Moreover, in simulations the questioning is not on a personal level but on a role level, and this is an important distinction. The duties inherent in roles usually embolden the low achievers and the shy.

MATERIALS

Some simulations require additional materials. These can range from paper and pencils to computers. They are usually listed in the organizer's notes and are the sort of items which are readily available. However, there are often other materials which are not listed but which can be added to increase the realism or add to the effectiveness of different aspects of the simulation — a vase of flowers, an in-tray, a ten-minute break in the canteen for coffee, large sheets of card and felt-tipped pens. This point is dealt with more fully in the next chapter.

ROOM SPACE

Room space is important in those simulations where several groups are operating simultaneously, particularly if it is desirable that they should meet in privacy (or secrecy). International affairs simulations and competitive business simulations are cases in point where an additional room or two can remove all temptations to observe or spy on what the other groups are doing.

However, if the simulation is right for the participants but only one room is available, then it is still possible to manipulate the geography of the furniture so as to lessen the chances of espionage.

Objectives

Just as authors sometimes design simulations, then find out what they have created and add the objectives on afterwards to fit the likely achievements, teachers might like to consider doing something similar.

This suggestion may sound like educational heresy, but it does have certain practical advantages for the teacher.

First, it means that the teacher's assessments and objectives are firmly based on observation — on seeing what actually happens. This is important in simulations, since what happens is often unexpected and may pass unobserved if the teacher is firmly concentrating on some predetermined objective.

Secondly, it is an attitude of mind which can broaden the spectrum of objectives. Instead of sticking to the obvious, it may encourage the teacher to look for other objectives as well and to evaluate all sorts of incidentals, which may seem peripheral but could be important, particularly to the participants involved.

Thirdly, it reduces the danger that the teacher will be conned by a high-sounding list of objectives contained in the publicity material for the simulation. Simulations have to work well to be effective, and objectives are not achievements. A good simulation with no objectives attached is preferable to a poor simulation with thousand well-meaning aims.

Fourthly, the sort of additional objectives that are likely to arise from actual observation of what is going on in a simulation are probably those asociated with education rather than training. Training objectives are easier to recognize and are more easily assessed. But educational objectives — the abilities and skills which make up so much of adult life — are not easy to recognize and may be impossible to quantify. But that does not mean they are not important. Confidence, for example, is tremendously important — so are organizational skills, the ability to use language and communicate effectively, and all sorts of personal traits and attitudes which help people to get on with one another.

In a simulation, important changes in a person's thinking and behaviour can occur and yet pass unnoticed because the teacher is concentrating on whether, for example, the participants are appreciating the problems involved in the use of land resources.

The argument is not that objectives are unimportant, rather the reverse. Objectives are so important they should be constantly appraised in the light of observation and experience. If a teacher finds that a particular simulation is consistently successful in achieving some unexpected yet desirable learning behaviour, this can be added to the list of objectives. If expected attainments do not occur, these can be dropped from the objectives. However, apart from choosing a suitable simulation and presenting it effectively, there is nothing the teacher can do to ensure that the objectives are achieved; that depends on the participants and whether the objectives are appropriate.

It can be useful, therefore, to examine some of the objectives which authors have listed as important, and see how these may influence a teacher's choice of simulations.

FACTS

Many situations, perhaps most simulations, are aimed solely at conveying facts and insights about a particular subject. Some of these are training simulations, designed to familiarize the participants with the subject and the procedures for dealing with it.

Some are tailor-made simulations for a specific company or organization and are intended to give the students a feeling of what it is like to be doing a specific job. These, however, tend to be guided exercises rather than genuine simulations, and in any case there is little question of choice involved as they are often so specialized as to allow no adequate substitute.

Leaving aside tailor-made simulations, the advantage of using simulations to convey the facts is that a more personal and committed form of learning may result. This supplements but does not replace the textbook. In addition, such factual simulations are easy to justify in the classroom and staffroom, as they are likely to be directly relevant to forthcoming examinations.

But of all the objectives which a teacher may have, the conveyance of facts about a specific subject is the most limiting as far as making a choice is concerned. If the teacher wishes to introduce a simulation about a specific event or historical development — the Wars of the Roses, for example, or the growth of railways in Africa — there may be few if any simulations to choose from. As there is now a rapidly growing number of simulations being published, however, it is worthwhile for teachers to delve into directories of simulations and publishers' lists to see that is currently available. As with any other form of learning materials, it is up to the teacher to assess the quality of the facts provided — their accuracy, their significance, their presentation — in the usual way. The teacher must exercise professional judgement and take into account the reputation of the author and publisher.

If the facts are contemporary, not historical, the teacher faces an additional problem — the facts may be (or may become) out of date. This is particularly the case in simulations like TENEMENT which deal with the specific and current social services, agencies, benefits, laws and regulations. Since such simulations convey an important feeling of immediacy and relevance in the areas of personal issues and general problems, students may well conclude that this is how things actually are in the details about pensions, job security and housing. The teacher can look at the publication date and try to update the materials or can explain that certain facts are no longer true; alternatively it can be presented as an historical simulation.

Choosing a simulation for its facts is not the same as choosing a textbook for its facts. In a textbook, the facts are there to be learnt; in a simulation, the facts are there to be used. If in a simulation the participants decide that some of the facts are irrelevant to what they are trying to do, they may dismiss these facts with no more than a quick glance. If the simulation has a huge scenario, with page after page of background facts which have little relevance to the action, not only may the facts go unlearned, but a feeling of annoyance may arise which has a negative effect on fact learning.

If, on the other hand, most of the facts presented in the simulation are potential weapons in a conflict of interests, they are likely to be scrutinized far more closely than would ever occur in reading a texbook. As well as learning relevant facts, the participant would be developing valuable skills of selection, analysis and presentation.

MODELS

In simulation literature the word model does not imply miniaturization, in which all the details are reproduced but on a small scale. When talking about simulations, the word model is used to cover the essential working elements of something — an economy, a political system, a society.

If the teacher's objective is a model, the choice is far wider than if the objective is to convey specific facts, because models do not have to be based on actual facts. Sometimes fiction and fantasy can provide a clearer picture of the way things work than an attempted simplification using actual organizations and specific examples. Fiction can be manipulated by the author to highlight essential aspects in a way which may be difficult to achieve with actual real-world components with a lot of historical and perhaps irrelevant clutter.

There is no right answer to the question of whether to choose simulations with actual names, organizations and places, or whether to choose simulations using fictitious elements. The familiar has the advantage of being easily identifiable; the unfamiliar has the advantage of giving the students an opportunity to take a fresh look at an old problem.

A model based on fiction may have to go into detail about procedures and background simulation 'facts' which might be unnecessary if actual organizations were used. On the other hand, a model based on actual names and places may run into the problem of which 'facts' are real, and can bias the outcome of the simulation in favour of those students who happen to know the most about that specific organization, country, or period.

DECISION-MAKING

Decision-making covers a wide variety of skills and disciplines. It is not the same as decision-giving; it is not just saying 'yes' or 'no'. Making decisions involves searching around for the most suitable decision, analysing the situation, and constructing hypotheses about what might follow certain decisions.

One of the problems of training is that people are trained to solve the same sort of problems and make the same sort of decisions again and again. This is why many courses in business management deliberately seek out unusual problems in order to broaden horizons and shake up established thought patterns.

So, in having decision-making as an objective in choosing simulations, the key question is the sort of decisions that are required. Are they specific, or are they general? Should the decisions be unfamiliar in order to give students practice in coping with the unfamiliar, or should they be restricted to previous decision-making experiences?

An advantage of unfamiliar decision-making is that it changes the existing classroom hierarchy which is usually dependent on factual knowledge. With new problems students have more equality of opportunity. For example, in STARPOWER, the squares are faced with a completely new problem — how to alter the rules. The situation has never arisen for them before. They cannot use previous knowledge. Usually they would have maps or some knowledge of what is one day's walk away but not in this case.

In DART AVIATION LTD decisions are made on familiar ground. Many participants will already have had experience of the problem of manufacturing paper darts. Decisions on prices, advertising expenditure and aircraft manufacture in the simulation are the same as they would be with a thousand other products. The decision-making is not a matter of life or death, but of gradual adjustments, trial and error, hypotheses, and analysis of results which can be thought of entirely in numbers. Ethics are not involved. The problem is related to maximizing profits, and this is true of probably most business management simulations based on a 'model' — an arithmetical model for determining the results of business decisions. If the participants have already had experience of this type of problem and this sort of decision-making, DART AVIATION LTD will help reinforce the learning. If, on the other hand, the participants are new to business simulations, the experience will be an extension of their personal experiences of buying and selling and advertising.

Again, there is no 'right' answer about whether to choose the familiar or the unfamiliar problem for decision-making. The familiar may be more related to training and the particular; the unfamiliar may be more related to education and the general. But circumstances alter cases and it is up to the teacher to select the most suitable objectives.

The same consideration applies to open and closed simulations. In arithmetical-based business simulations, the arithmetic is an objective criterion for measuring the success of the decision-making. The problem is to crack the code and this is done by analysing the result of round-by-round decisions.

In closed simulations there may be one right answer or several, but there is an objective criterion which enables students to say 'Oh, yes, I see, we got it right (wrong).' Open-ended simulations, on the other hand, may not only lack any objective criteria, but may also be deliberately designed to provide simulation 'facts' of equal weight to conflicting interests to balance arguments and make everything a matter of opinion. Behavioural and human relations simulations tend to be open-ended; so do conflict simulations in local government or international affairs. In open-ended simulations the decision-making is not a neat and tidy scientific affair. Thus, STARPOWER is unfamiliar and open, whereas SPACE CRASH is unfamiliar and closed. STARPOWER is unpredictable — almost anything can happen — and who is to say whether what happens is right or wrong? In SPACE CRASH the participants either live or die and almost always die.

However, in a simulation, unlike an examination, the result is rarely important. What matters is how the result is arrived at. Secondary school pupils tend to prefer closed simulations because they satisfy a 'What's the answer?' approach to learning. Whether this is a good enough reason for a secondary school teacher to select closed simulations is another matter. The teacher may take the view that one objective is to try to persuade the pupils out of the habit of asking 'What's the answer?' to anything and everything. In many areas of life 'What's the answer?' is not an appropriate question. Questions themselves can be more important than answers. Exploration can be preferable to arrival. Decision-making can concern opinion as well as fact.

COMMUNICATION SKILLS

Through effective communication, we deliver our ideas. It is useful to be able to communicate and even better to be able to communicate effectively. It is inefficient for brilliant ideas to be concealed by inarticulate mumbling.

One of the great advantages of simulations is that they are self-activating and provide scope for a far wider range of useful communications than normally occurs in education. Instead of the student communicating with the teacher, usually answering questions to demonstrate that something has been learned, in a simulation the student learns *by* communicating

It is also useful to remember that emotions and motives can be important elements. In the five simulations described in the preceding chapter the communication is motivated by feelings — about poverty, power, flight, life, love.

Simulations in the same subject field can often cover a wide range of communication skills — diplomacy, arguing, interviewing, reporting, note taking, drafting, editing, organizing, presenting a case, speaking in public, listening. In many schools pupils go in at one end and come out at the other without any personal contact with many of these skills. Practice brings skills, and skills bring confidence; without practice there is no confidence. Many students hate the idea of speaking in public and would run a mile if asked to take the chair or to present a case. Yet these skills have a wide transferability. The skill of diplomacy is of value to all people, not just diplomats. The ability and confidence to speak in public are of value to others who are not public speakers. Being able to speak in public does not mean that one has to do so; but it can be most satisfying to know that one could cope if the occasion arose. Obversely, lack of practice, lack of skills and lack of confidence can on occasion result in near mental paralysis if a person is afraid of being called on to speak.

If practice in communication skills is one of the objectives of the teacher in choosing simulations, it is probably a good idea to seek out simple, argumentative and emotional simulations as starters. Some simulations are specifically designed to encourage communication skills and all simulations involve communication. So it may be worthwhile looking at the difficult part of a simulation — the bit that is left out — to see what sort of communication is involved. If there is a series of simulations on the same subject, this could be a gradual way of giving communication practice. Alternatively it may be better to look for a wider variety of roles and situations to cover more skills.

Fortunately, the specific subject is not all that important since it is difficult to be clear and articulate in one subject without also being good at communicating other subjects too. Preparing one speech helps to prepare the next one, even if the subject is different. Diplomacy in international simulations may help diplomacy in family and personal relations.

LANGUAGE SKILLS

Language skills are allied to communication skills, and sometimes the two phrases are used interchangeably. In practice, language skills are usually taken to refer to the sort of skills taught by the English department or by teachers of English as a foreign language and these are not identical.

Yet, in both fields, the first problem may be to encourage the students to talk. One teacher in a north London secondary school observing his class take part in their first simulation, said with astonishment, 'I've never heard half of them speak before.'

It is not just a question of language, but a question of confidence. Simulations have the advantage of removing the teacher, who is sometimes an inhibitive figure, respected yet feared. To the shy student it may be better to keep quiet rather than say something wrong and be laughed at.

The British Council is in the forefront of advocates of simulations. Kerr (1977) says:

> They ensure that communication is purposeful (in contrast to the inescapable artificiality of so many traditional exercises and drills); and, secondly, they require an integrative use of language in which communicating one's meaning takes proper precedence over the mere elements of language learning (grammar and pronunciation).

If language is high on the teacher's list of objectives, subject matter is probably a good deal lower down. It may be a prescription for failure to set out with the aim of matching the students' interests with the subject matter of the simulation. If the teacher has a group of Spanish engineers as students, it is courting disaster to place a blind order for a simulation that is about (a) Spain, (b) engineering, or even (c) Spanish engineering. The scenario may be bulky, the language content may be slight, and the interaction virtually non-existent. The main language skills may come afterwards when the participants complain that 'It is not like that' in their branch of engineering or in their part of Spain.

BEHAVIOUR

Quite a number of teachers introduce simulations for exclusively behavioural objectives and these can vary considerably.

One is the straightforward objective of improving the behaviour of the class. If there have been antagonisms, frustrations and tension among pupils, boredom or dissatisfaction with the course, or potential hostility between pupils and the teacher, a good simulation can work wonders. It removes boredom, redirects the activity, and extricates the teacher from a confrontation.

Other behavioural objectives can range from self-awareness to an examination of the hidden motives and attitudes in society. By their nature such simulations tend to be controversial on all sorts of levels. For example, Zuckerman (1973) notes that most racial attitude simulations ask the participants to get into the black experience. He says this encourages 'shallow depth responses along the line of "Lo, the poor black man, who is so mistreated by those *other* people." ' Shirts (1970) says that simulations about the black community are generally written by people from the suburbs and are based on a series of unfounded clichés about what it is like to be black, which not only encourages stereotyping but creates an attitude of condescension towards blacks. Also, says Shirts, such simulations can give the students the impression that, having taken part in the simulation experience, they know what it is actually like to be discriminated against or what it is like to be black.

These criticisms are not altogether satisfactory. For example, if there is stereotyping, this suggests that it is not a simulation but an informal drama. If a role card for a member of the housing department says, 'Generally does not favour allocating houses to blacks in white areas', this is personality imprinting — a stereotyping which denies the participant the right to make up his own mind in the light of the situation. Similarly with the question of condescension, unless it is role-play in which the person is asked to behave in a condescending manner, condescension is no more predictable than sympathy or respect. In a genuine simulation, the participants can feel any way they like, providing they do their job as best they can.

Behavioural simulations, like textbooks or television programmes, may or may not be typical of specific problems. But no medium exists in isolation and any individual example in any medium is unlikely to convince a student worth his salt that he 'knows it'.

In addition to the 'big issues' there are plenty of what might be called 'local' behavioural simulations. Interview simulations are a case in point. The objective is to help the participants behave more effectively in interviews.

There are personal conflict behavioural simulations — conflicts between boss and employee, between parent and child, between teacher and pupil. Some of these are close to role-play exercises, but others are genuine simulations involving groups of people, documents and decision-making. The dividing line is not so much the number of participants or the documentation, but whether it is teacher-guided, whether it is personality-imprinted, or whether it is based on job function.

FRIENDSHIP

Friendships develop in simulations and there is no reason why they should not be added to a list of objectives. Indeed, some teachers use a simulation for no other purpose than to help students get to know each other at the beginning of the academic year. Not only does it help students get to know each other, it also helps the teacher to get to know the students in a way which is different from restrictive teacher-student orientated behaviour. With simulations it is quite common for episodes to occur which take the teacher by surprise. 'I had no idea that Mandy had it in her' is a common remark. Consequently, the teacher has learned something, and so probably has Mandy and the other students.

Friendships involve more than just friendly feelings; they imply understanding and communication, and working together effectively. All these can be very useful at the beginning of a course, and pay dividends later on in work which might be completely dissimilar to the subject matter of the simulation. Teachers may tend to undervalue friendships, thinking them personal, not educational, but they are very important in education just as they are in the world outside the classroom.

PREDICTION

This category of objective has to be mentioned since it is extremely valuable to certain specialists who wish to find out in advance how certain things are likely to work.

Foreign ministries may want to know 'What is likely to happen if we did so and so?' A simulation is the best method of finding out, short of actually doing so and so. A defence ministry may wish to know the most effective strategy for some newly-developed weapon — but it cannot start a war to find out. There is no alternative but a simulation. Local authorities may wish to know if their resources for dealing with a disaster are likely to be effective. Again, it is a matter of simulating the situation. But prediction is not an objective that the practising teacher need consider. If it were worth considering, then almost certainly the teacher would already be using predictive simulations.

SELECTING

The bulk of the work of selection has been done once a teacher has assessed the resources available and has settled on the objectives. All that remains is the practical step of examining and trying out potentially useful simulations.

The additional points to watch for have already been dealt with in earlier chapters — the need to look at the design of the simulation, to see if it is fully participative, if it is well balanced, if it is provocative, stimulating, interesting, emotional, involving, and so on.

If the selection of the simulation is appropriate then the answer to whether the simulation will work effectively depends on the participants and the skills of the organizer. These issues are dealt with in the next chapter which concentrates on simulations in action.

Chapter 5
Using simulations

Participation

If the teacher participated in a simulation at the choosing stage, all well and good. But if not — perhaps because the simulation had already been chosen and was awaiting the teacher's use — the teacher should arrange a participation session with a few friends or colleagues.

Some teachers protest at this advice, because they see it as a waste of time. They see the simulation in the same light as a book or film and think that all that needs to be done by way of preparation is to inspect it. If it is not possible to persuade or bully people into having a complete run-through, at least the teacher should make every effort to enrol a few people for participatory sampling, as suggested in the last chapter.

With something like an arithmetical model-based simulation the teacher should actually fill in a decision form and process it — perhaps with other group decisions — in order to arrive at the resultant number which comes out after the data have been fed through the arithmetical formula.

The sort of thing which can happen is that a group fills in forms wrongly without knowing it and so receives 'wrong' answers as a result. By the time the students have acquired enough experience of the simulation to realize that something is wrong, it may be too late to unscramble the mess. Everyone will have to start again or abandon the project — and either eventuality will do nothing for the teacher's reputation. Reading the instructions and saying 'I don't need to try it, I understand it' are famous last words.

Organizer

Although the participants 'own' a simulation this does not mean that the organizer's role is unimportant. On the contrary, the organizer has vital and varied functions, including observing and assessing.

The first job of a novice organizer is to shake off any inappropriate habits or thoughts derived from the experience of teaching or instructing. The main part of every simulation, the interaction, is not taught. By training and by habit teachers interrupt, guide, explain, give hints, smile, frown, and in many subtle ways (including silence) try to help students learn. But if a teacher tries to do this in a simulation, it stops being a simulation and becomes a pseudo-simulation or a guided exercise.

Even during the briefing and de-briefing it is useful if the relationship between organizer and students is that of the relationship between professionals — respectful, polite, slightly distant but enthusiastic about the speciality — in this case simulations.

Once into the action, the organizer should aim to be invisible, to merge into the background or to assume the protective colouring of a plausible role such as usher, messenger boy, furniture remover, or friend of the editor or managing director. When it becomes necessary to speak to the participants, it is most valuable for the organizer to adopt appropriate protocol. Instead of saying 'If any student...' or 'If any participant...', it is better to use the appropriate titles — 'If any Honourable Members...', 'If any executives...', 'If any councillors...', 'If any journalists...'

As mentioned earlier, the concepts of autonomy, consistency and plausibility are vital, and these provide guidelines for the organizer and for the participants. Reality for the participants is the interaction. It is theirs; they own it. Conversely, what happens (or is supposed to happen) outside the area of the simulation is the responsibility of the organizer, or the organizing team if it is a big simulation. This means that if any participant orders an event to happen outside the room — calls for a day of prayer, calls for troop mobilization, calls for a strike, calls for a protest march, calls for the police — it is up to the organizer to decide what happened and to inform the participants accordingly in whatever manner seems appropriate within the realism of the simulation.

Some simulations have their own built-in mechanism which makes this unnecessary. In management simulations, and within specified limits, certain orders are automatically carried out according to the mechanism. Providing the companies in DART AVIATION LTD fix their advertising costs within certain limits and in multiples of £1000, this advertising will occur. There is no chance of the advertising agency going bankrupt or the advertisements being declared illegal or ineffective; that is not part of the simulation. Nevertheless, should any group of participants wish to take decisions outside their area which are not covered by the 'rules', it is a matter for the organizer alone to decide what happens.

The participants have power within the simulation interaction; the organizer has power outside it. Power, authority, duty and responsibility are thus clearly defined and are clearly separated by the line between inside and outside.

Detachment

Within a good simulation organizers are often subjected to the strongest temptations to participate themselves. In the swirling action, the vigorous arguments, the emotional, imaginative and momentous events, the organizer can sometimes be observed grinning with excitement and holding back from rushing in to participate. Extraordinary though it may seem, the teacher sometimes does actually sweep forward and assume some role or other. This is bad intervention. There can be good intervention if, for example, a key role is obviously vacant. Perhaps there should be someone in the role of Secretary-General of the United Nations for a few moments, or a policeman, or a king, or an usher. These are temporary interventions to assist the mechanics, rather than interventions for personal reasons. It is a general and a safe rule never to interfere, nor to give any signs of pleasure, displeasure, surprise, boredom, annoyance, appreciation or exasperation.

A poker-face is virtually obligatory. It is not easy but is worth cultivating since it deters students from looking at the organizer for signs of acceptance or rejection. Poker-faces help to preserve participant power and responsibility. At first this may seem strange and unusual to the participants, but they soon see the value of it and take pride in 'owning' their simulation, which would be impossible if they had to keep looking over their shoulders at the organizer.

This detachment by the organizer could be explained in the briefing: 'My job is to remain poker-faced. There should be no point in your looking at me. Should you see me smiling or frowning, I am not doing my job properly. Try to take no notice of me at all. A simulation is not like a play; you are not going to be coached and guided and stage managed. We are not aiming for a perfect performance. You are on you own. You've got a job to do. Just do your best and ignore me.'

Observing

With some simulations the organizer has no problems about observing what is going on without interfering in the action and without disrupting the participation. If the simulation is a public debate, the organizer can sit poker-faced in a corner of the room and can see and hear just as much as any of the participants. But a problem arises when groups meet separately and in secret. Should the organizer pull up a chair and say, 'Don't pay any attention to me' and listen to what is being said? With three or four groups, the organizer could move about and enter the intimacy of the circle of confidential talk, and then move off and on to another group. Is this a good idea? Is it a good idea even if the organizer resists all temptation to smile, frown, look appreciative, perplexed, or whatever?

In general, the answer must be no, it is not a good idea. If it is a straight choice between ignorance and interference, it is better to remain ignorant rather than risk distraction or interruption or subtle manipulation.

This is because what happens happens, and the interaction should be the sole responsibility of the participants, without interference. It can, of course, be recollected and discussed in detail at the de-briefing. The group can say, 'We had a lot of trouble deciding whether to do so and so, and Mr X said such and such and Miss Y replied...' The fact that the organizer did not listen to the actual conversation is not important; what is important is that it took place.

This advice may be very difficult to take. Teachers like to know what is going on. If they know what is going on, they can use the information afterwards and give advice, hints and guidance in the de-briefing. It is all part of the teaching technique to observe and comment.

But a simulation is not teaching. It is learning and the learning will be placed in jeopardy by interference. The learning is not just the learning of facts; it is behavioural and concerns power and responsibility and should not be diminished. No teacher walks into a cabinet meeting or boardroom meeting and says, 'Don't mind me, I'm not here, just carry on normally.' If such a remark is to be made, it should be made during the briefing, not in the middle of the action.

There are, however, certain techniques for observing without interfering. One technique is to walk around slowly at an even pace, moving from group to group but avoiding eye contact. Another method is to sit at a 'blind' spot, perhaps in a corner of the room, or take a seat at the side if two groups are facing each other.

Role allocation

Teachers who are experienced in using informal drama and role-play exercises probably have their own favourite method of allocating roles. But a simulation is quite different from an exercise and an informal drama. Different considerations apply and what may be suitable for the one may be unsuitable for the other.

There are two kinds of roles — group roles and individual roles. These can be considered separately even though there are many simulations which contain both forms.

Group roles

It could be a group of business executives, government ministers, tribespeople, or whatever, but the key question is which group the individual participants should belong to, rather than what their function will be as members of that group.

One question is whether students A, B and C, who always work together, should be allocated to the same group. Another question arises if groups have important distinguishing characteristics which evoke preferences or prejudices. For example, who should belong to the Bosses' Party and who should be members of the Workers' Party? Role allocation can be random or by individual preferences, or by teacher selection. Each has advantages and disadvantages.

Teacher selection has the advantage of putting the decision-making into the hands of the existing classroom authority — the teacher — who should be best placed to make the selection on whatever criteria may appear appropriate. The teacher can keep together a group of friends or can split them up. Deliberate selection allows the teacher to restore the balance of arguments in a simulation by placing the more able students in that group which is the least popular among the students, eg the chemical company management rather than the anti-pollution protesters.

There are potential hazards in teacher selection of group roles. One is that it may arouse resentment among the students; they may feel manipulated or discriminated against. If so, this would be contrary to the basic principle of simulations which is to give as much power and responsibility to the participants as possible.

Another problem is that if the teacher and students are new to simulations, the criteria of student ability may be inappropriate, since it would be based on normal classroom behaviour and results. Surprising things can happen in simulations and the usual hierarchy can be rudely shaken.

Probably the easiest method of allocation to groups is to keep together those students who usually work or play together. This still leaves the problem of bosses or workers, landlords or peasants, hunters or farmers, traditionalists or revolutionaries, but it does mean that during the course of the simulation there is a minimum of friction within individual groups. Naturally, the strength of this arrrangement depends on the strength of the group feelings in the class.

The random method of group allocation offers a solution to several problems simultaneously. First, it is fair and can be seen to be fair. It avoids allegations of favouritism. It helps the students to assume the responsibilities and duties of professionals, and it helps to establish the teacher in the neutral roles of organizer and observer.

The main disadvantage of the random method is the resentment felt by friends who find themselves separated. You could explain however, that this is one of the hazards of life, and that it is a good idea to try to co-operate with people who may not be friends and may sometimes be enemies. This disadvantage is likely to diminish as students widen their circle of acquaintances and friends, which could turn a short-term disadvantage into a long-term gain.

It is worth making the point that random selection really should be seen to be random and not an announcement, 'I'm randomly allocating student A to the king's party and student B to the slaves' compound.'

Some simulations can be re-run with a change of group roles in which case role favouritism is no longer an issue, and this leaves only the problem of the composition of the teams.

Individual roles

An individual role means that an individual has some particular function, responsibility or knowledge which is different from the other participants.

In TENEMENT, although there is a group of seven tenants, they are all individuals with individual role cards, individual problems, and individual circumstances. The people in charge of the agencies have no personal or individual roles, just job roles. STARPOWER has social group roles only, and DART AVIATION LTD has job group roles only. SPACE CRASH has individually named role cards but there are no differences of job, age, social class or gender. The purpose of these role cards is to convey separate pieces of information and advice. THE LINGUAN PRIZE FOR LITERATURE has no role cards, and uses memos to give information and advice to the different groups. The participants tend to use their real names — 'Hello, and welcome to Arts at Noon. My name is Sara and in the studio we have two judges, Mr Ricco Brown and Miss Amanda Patel, and also Mr O'Hare who is an editor. They will be talking to Sal about the literature prize. Over to you Sal...'

For the organizer, a key problem is what to do if a simulation has a role or roles implying specific skills, such as a technical expert. An important consideration is whether the scientific data are available to all the participants, perhaps in a 'library', or are restricted to the role card or documents belonging to the expert. It may also be that the public documents contain only general scientific evidence while the individual role card and private documents give additional evidence in detail.

Other considerations are whether the scientific evidence is disputed. Are there two or more 'expert' views, or does the simulation provide only one undisputed view? How important is the scientific evidence in a given simulation? Are the overriding factors public opinion, financial priorities or statistics about acceptable or unacceptable levels of pollution?

Answers to these questions will help in deciding how to allocate the role of expert. If individual roles are allocated at random and student X or Y or Z gets the role of expert, does that unbalance the simulation or render it less effective? Or are there safeguards? Can other participants unearth the information and question the testimony of the expert? Does it matter much anyway since the educational aim is likely to be practice, not perfection?

Cases should determine decisions. But it is probably a good idea, in those simulations where the role of expert is of some prominence, to allocate the role at random but to allocate two participants to the role. They will back each other up and it is not unrealistic for an expert to have a colleague or assistant. This technique can also be used in simulations which require the role of chairperson, whether in the national assembly or the public inquiry, or a sub-committee of the local boys' club. Even if all the participants are themselves top-level adult experts, it might still be a good idea to consider whether it would be desirable to allocate a second person to a role — perhaps deputy speaker, deputy leader, deputy chairperson, deputy prime minister.

The arguments in favour of randomness in individual role allocation are similar to those already given in the allocation of groups. The technique is fair and is seen to be fair. People can develop by having responsibility, and the aim is to give opportunities, not to engineer perfection. Random allocation enhances the principle of participant power, and at the same time it strengthens the role of the organizer by emphasizing the organizer's impartiality, thus avoiding the damaging accusation of role manipulation. As with group allocations, the main disadvantage of randomness in individual allocations is likely to be in disrupting the pattern of friendships or working acquaintances. But this is likely to be temporary and may well have longer-term benefits. The most difficult simulation for random role selection is always likely to be the first simulation; afterwards it becomes much easier.

An alternative to random choice and teacher selection is to allow the students to volunteer for specific roles. In practice, this is not usually a good idea because of possible disputes or feelings of unfairness. It opens the door to the extroverts to seize the best (or easiest) roles. Resentment about this may seethe beneath the surface and can damage the effective running of the event.

Presentation of materials

What should be a simple matter of presenting the right materials at the right time seems to cause far more trouble than it should. Possibly one explanation is that the teacher has not participated in the event and fails to appreciate the importance of sorting out the documents before venturing into the classroom — perhaps using colour coding, envelopes, paper clips. It may help, for example to have the notes for participants photocopied on non-white paper.

The need for care in presenting the materials is not simply because it is more efficient to do it properly. Another reason is that in a simulation the penalties for mistakes are likely to be greater than in traditional teacher-student situations. If, in normal classroom discussion, the organizer has to fumble around to find the right bits of paper, this may be rather annoying or amusing to the students, and they probably do not object to the delay. But in a simulation, which tends to develop a considerable dynamic quality of its own with a great deal of involvement, any delay caused by organizer inefficiency may well result in considerable exasperation from the participants. Furthermore, the mechanical breakdown may jerk the participants out of their roles, and it may take them some time to re-establish the flow and realism of the simulation.

The actual loss of a document can unbalance the simulation to such an extent as to make it virtually unworkable. In ordinary classroom teaching, the loss of one fact-sheet out of ten may not be disastrous, but in a simulation, where the pieces fit together and are often interdependent, the loss of one part can easily nullify the whole. So the organizer must allow sufficient time to see that all the required documents are available and that they are arranged in the right order for presentation.

In preparing and arranging the documents, the organizer should consider not only the action part of the simulation, but also the briefing. Are any additional documents desirable in order to clarify the basics of the simulation? For example, if there are no notes for participants, it may be a good idea to prepare some — perhaps half a dozen sentences outlining the main points. Not only will this inform the participants, it will also offer a safeguard in case the organizer omits some vital information in the verbal briefing. It avoids the organizer being placed in the embarrassing position of having to interrupt the simulation and say, 'Sorry, I forgot to tell you that...'

If the organizer does not prepare or present any notes for participants, it is useful, possibly essential, to prepare a checklist of points to make during the briefing. Various home-made documents can also help to explain to the students how the mechanics of the simulation will operate. The organizer might draw up a chart or diagram or map showing time sequences, location of furniture, and so on.

Other types of materials and equipment can help to make the simulation more efficient or more plausible, depending on the situation. Here are a few examples: clean note pads and sharpened pencils, little flags, name tags, labels for locations, word processors, typewriters, telephones, pocket calculators, vase of flowers, clock, coffee and biscuits, wastepaper baskets, formal message pads, tape recorder, video camera (for media simulation), clip boards, maps, tray of paper clips, carafe and glasses of water.

Classroom furniture

The question of the geography of the classroom furniture does need to be thought out beforehand. Provided the teacher has already participated in the simulation, there are no real problems about where to put the tables and chairs, but plenty of opportunities to add to the realism.

The starting point can only be the simulation itself — its nature and structure. The requirements can be one single table around which everyone sits or separate tables or areas for different teams or individuals.

If secrecy is important, teams should be as far apart as possible — perhaps one in each corner of the room, or with different rooms of their own, as recommended in some business and foreign affairs simulations.

Some simulations change the pattern of interrelations at different stages, and in this case it is important that the classroom furniture should be changed as well. It may seem a chore to move a few tables around, but it is always useful and sometimes essential that this should be done.

A team should have a base from which to work. This should be allocated by the organizer to a suitable spot, bearing in mind the location of other team areas, the organizer's own area, and the likely traffic between bases.

Should the simulation involve interviews, furniture arrangements should copy the normal furniture arrangements for interviews. In a public inquiry the chairperson and advisers should have one table dominating the room with the other chairs either facing this table in rows or as cross benches if there are 'for' and 'against' parties.

What should not happen is that the teacher walks into a class in which there are neat rows of desks, simply hands out material and asks the students to form teams and find tables for themselves. This is a prescription for time-wasting and muddle. If some teams should be close to each other, the organizer must arrange this; if far part, this also should be indicated. Maps of the classroom showing which groups sit where at each stage of the simulation can be useful. Such classroom plans not only save time but allow the students to do their own furniture moving.

If the simulation involves a map of a geographical area on which countries (teams) are represented, the classroom geography might be copied from the map, with neighbouring countries having neighbouring teams. This may cause an espionage problem if the neighbours are hostile. There are several ways of solving this: a screen between the two teams, a strong warning against 'cheating', espionage not being allowed, or the two teams sitting facing away from each other.

The location of the organizer's desk should be within easy access of all the participants, but preferably not impeding busy traffic routes.

If the organizer has to fumble uncertainly with classroom furniture during the course of the simulation, this is almost as bad as not getting the materials right. The organizer who says, 'Well, perhaps we might put this desk here, and you could go over there — you, not you — and then we'll have those materials over here on this table' can disturb the concentration of participants who are in the middle of an exciting and involving experience.

Special provision will have to be made if the simulation includes the media — newspaper, radio or television. Each media organization should have its own base with the necessary equipment.

A studio can be a problem for teachers who are not experienced in this sort of thing. However, a simple and acceptable solution is a tape recorder with a microphone placed on a chair on a table. A coat over the back of the chair placed between the microphone and any likely source of external noise helps to improve the quality of the recording.

Broadcasts should always be live. (Fancy recorded inserts rarely work and take away the personal quality of immediacy.) They should also be done standing up, moving towards and away from the microphone at appropriate moments if more than two people are involved. The alternative of sitting at a table usually introduces a mass of clunks and squeaks as the microphone is pulled in front of the speaker, or chairs are thrust back and forth.

Television broadcasts are much the same as radio broadcasts, except that the viewers should all be at one side of the room and there should be markers to indicate the left and right of the screen to indicate when a speaker moves into or out of camera shot. So all-pervasive is television that this convention is readily accepted by the students. No cameras or technicians are needed.

Masterman (1980, 1985) gives some useful examples of how simulations can help in courses concerned with learning about television. Media simulations can be done in style with actual video equipment and with trained and practised operators.

However, a warning note should be sounded about using available electronics. Many a simulation has been brought to a shuddering halt by failure to operate the equipment correctly. Incidents causing laughter can easily arise through failure to spool tape successfully, cue participants, or edit the videotape. The simulation can grind to an embarrassing halt, rather like a lecture where no one knows how to operate the projector.

If done successfully, of course, there is the advantage of an electronic record of what happened which can be used for demonstration purposes later. But the organizer should think carefully before embarking on the use of electronics. Is it necessary? Is it desirable? Is it merely a gimmick? Is it worth risking the value of the simulation? Will the actual process of recording the action detract from the realism? With a news conference, of course, it is permissible to accept that it will be recorded and that things could go wrong in setting up the lighting, microphones, and cameras. But if it is a confidential cabinet meeting, or an interview of an applicant for the headship of a comprehensive school, a battery of electronics — and the possibility of hold-ups and failure — may not be worth the effort, and would almost certainly detract from the realism and inhibit the action.

Business and military simulations can be different. If these need computers to work out results, they must be provided.

Timing

The timing of simulations can present problems or involve guesswork as some activities can be completed quicker than expected or continue longer than expected. Again, the teacher's own participatory experience in the simulation is invaluable, not only in assessing how long things will take, but also in drawing up contingency plans in case the timing does not work out.

However, nothing really disastrous can go wrong, provided that no vital activity is started without sufficient time to complete it. If the climax of the simulation is a radio transmission, it will be a shambles if the lunch bell goes when the broadcast is only half completed. Similarly, if a person is to make a set speech, it should not be begun if it cannot reasonably be finished in the time available.

If the simulation contains an activity which has to be treated as a whole and not interrupted, then the timing must be tailored accordingly. Deadlines must be firmly laid down, guillotine procedures established and starting times observed. Deadlines, in particular, must be regarded as sacred. The participants have the role of professionals and they must behave like professionals. If the broadcast is due to start at 2 pm, the participants cannot appeal to the organizer to 'Give us a few more minutes, we are not quite ready.'

Breaks in simulation activity should not be regarded as necessarily undesirable. They give the participants time to think out their strategies and explore their options. A chat over a cup of coffee or an evening at home can help the participant ease into the role.

In deciding how to fit the simulation into the time available, consideration should be given to the breaks. It is helpful if the briefing, and perhaps the handing out of some materials or roles, takes place on the day before the action, although contingency planning may be needed in case someone does not turn up or unbriefed students arrive.

Some simulations last for days, weeks or even months. In these cases it is sensible to make sure that teams have adequate resources to carry on in the case of unavoidable absences.

Stages

A simulation which has various stages may present problems in moving smoothly from one stage to the next.

With model-based business simulations each stage may be similar in form to the last one and no great changes need occur. Nevertheless, it is worth considering what should happen when one team finishes its decision-making much earlier than the rest. Do they analyse the data further, do some other task, chat, do crossword puzzles, go for coffee, or snoop on the other teams? The organizer should have some suggestions to put forward in the briefing, and this means thinking about the problem beforehand. If there is a delay in processing the decision-making forms in order to present the results to the teams, it is useful to warn the participants in the briefing, thus avoiding or reducing any unnecessary frustrations during the action. However, there are various devices for reducing or even abolishing such delays. One technique is to time the end of a session to coincide with a normal classroom break — leaving the organizer (or clerk) to feed the decisions into the arithmetical model and produce the results in time for the students' return.

Although some business-type simulations still require the organizer to work out the arithmetic, the advent of pocket calculators has reduced the delays, and if the simulation is computer-assisted then delays are virtually abolished. In those simulations in which the stages are not based on statistical models a break may nevertheless be valuable. Suppose, for example, a simulation includes events which have to be reported in written form, as when political simulations include newspaper journalism. In such cases it could be useful to end the news conference shortly before a break and for the journalists to start writing their news stories immediately after the break. The break allows them time to think how they will present their stories.

Another alternative, which happens in real journalism, is for any long event (conference, debate, public inquiry) to be reported by a team of two or three journalists who take it in turns to take notes and then write their stories.

Computer-assisted simulations

As mentioned in the last section, computers can greatly assist model type simulations by rapidly producing statistics which show the result of decisions made by the participants, or which represent changes in the simulated environment — the weather, the number of hospital beds available, the height of floodwater.

If a computer programe is an inherent feature of a simulation then the important issues arising from this will doubtless be covered by the organizer's notes. This leaves the question of whether computers should be used when they are not a built-in feature, and the answer depends on circumstances and aims.

If there is sufficient time and computer expertise, then the organizer and/or students can produce their own programs for a simulation which can be of the number-crunching variety, or a data bank. Such programs could be written for simulations like TENEMENT and used by the agencies. If the aim is to model local housing regulations, or unemployment benefits, or list community services, then such a program could be written, and once written it would be relatively easy to increase the information and update it when necessary.

Equally appropriate would be the use of computers as word processors in THE LINGUAN PRIZE FOR LITERATURE. Any simulation involving journalism or authorship could be helped by word processing. If the software includes facilities for printing large letters and various graphics, then these can enhance the production within a simulation of a page of a newspaper, an official announcement, a company's logo, an advertisement.

Perhaps the main danger of a computer-assisted simulation is that it could reduce human interaction. If one participant operates the keyboard and takes decisions without proper consultation then the other participants would be reduced to passenger status. Instead of talk and thought there could be silence and dissatisfaction.

Another disadvantage is that some computer operations take more time than is justified by the results, and may be inappropriate, implausible, and inefficient. To ask participants to use computer graphics to design a plane in DART AVIATION LTD or to design a fashion theme in THE RAG TRADE would be to dispense with paper, scissors, glue, coloured pens and much of the interactive satisfaction.

Despite these dangers it is certainly worth looking at the options and seeing whether computers can be used effectively.

However, this is saying no more than the general piece of advice — when preparing to run simulations it is valuable to explore the possibilities and potentialities of any appropriate equipment, materials and facilities. These could include typewriters, video cameras, tape recorders, large sheets of coloured paper, overhead projectors, dictionaries, pocket calculators, the help of the art department or the school secretary, the help of an outside expert, magazines, bottle tops, scrap paper, name tags, clip boards, elastic bands.

Reruns

The possibility of a rerun should be considered before the briefing because it might affect the way the simulation is presented, and also the way the simulation operates. For example, if the students know that there will be, or may be, a rerun, they will have an additional motive for paying attention to what the other participants do, as they themselves may have that role in the rerun.

Reruns allow the participants to switch roles and find out what it is like being on the other side of the table or the other side of the argument. They enable the participants to have another go. This is important, as students are usually self-critical of their behaviour in simulations, and are often far from satisfied with their actions. Knowing more about the simulation at the end than they did at the beginning, they are often anxious to try it again.

For the organizer, the scheduling of a rerun means that there need be no anxiety that certain 'lessons' or opportunities are missed first time through.

However, no simulation should be rerun unless there is a good reason for it. Simulations which have repetitive stages are unlikely candidates for a rerun, unless the arithmetical model or some fundamental condition is changed. Simulations with hidden agendas are unworkable the second time, because the 'answer' was revealed the first time. Equally unsuitable are simulations which are like puzzles with one right answer.

Adapting

Many teachers adapt materials. Making adjustments to suit the classroom conditions is permissible, but it should never be wholesale adaptation and no part of the author's work should ever be copied or reproduced without permission as this is illegal, being an infringement of the copyright laws. In music, for example, taking a few bars of an essential melody may constitute an infringement of copyright. As far as the adaptation of simulations is concerned, the main danger is that the teacher will start the adaptation before the simulation has even been tried out.

Because simulations are so difficult to assess in their packages, it is often the case that the teacher decides that a certain document is too complicated or that a role is unnecessary, or that a lot of extra facts have to be fed in before the participants can start. The teacher can ruin an excellent simulation in this way. Key bits which are parts of the checks and balances can be removed or rendered ineffective. For example, the simulation may contain a boring, jargon-filled document about health hazards or unemployment benefits, and the teacher may decide not to hand it out. Yet the simulation within itself may generate a need to tackle such a document for some specific purpose — to report on it, to interview someone about it, to use it in an argument as evidence, to ask for its amendment, etc. In the action part of simulations, documents are rarely read dispassionately.

Generally speaking, the teachers who are experienced in using simulations are more willing than inexperienced teachers to run a simulation according to the recommendations of the author — at least for the first run through. This means that any subsequent adaptations will then be based on experience, not guesswork.

For example, what can be done if the simulation works reasonably well but has a bulky scenario which has to be digested before the action starts? There are several possibilities. The information might be fed into the simulation in stages. Some of it might be incorporated into the individual role cards. The mode of presentation can be varied — a letter in students' pigeon-holes, a tape recording, even a telephone call. If the organizer sifts through the information, some of it might be separated into a 'library' where it can be delved into and skimmed through according to the various strategies and interests of the individual participants.

Should the simulation seem to stick at particular points, with no one showing initiative, the organizer can see if it needs a drop of oil — perhaps the introduction of the role of journalist, or of some provocative document.

Similarly, the simulation may seem to be less effective than it might be because of part-time or passive roles. For instance, can a jury be dispensed with? Is it necessary to have a United Nations Secretary-General — or could the organizer take on this role on odd occasions? Is a banker an essential role or could it be left to the participants to manage their own transactions?

Practice and experience are necessary for distinguishing useful waiting from boring or frustrating waiting. A person waiting to appear before an appointments board may be sitting around and, to all appearances, doing nothing. Yet mentally the participant may be very active in rehearsing strategies and predicting likely questions. Waiting can be alert, watchful and strategic, with the potentiality for action, and the mental recording of who is doing what and why. This is not the same as looking blankly out of the window, reading a comic, doing a crossword, or telling funny stories.

There can be several reasons for adding to a simulation. The addition of 'public opinion' — perhaps journalists — to a foreign affairs simulation can add more realism, introduce a constraint to the use of power by leaders, and provide an impetus to diplomatic moves. The organizer can add facts — perhaps in the form of leaflets in a 'library' — in order to give more background data and ammunition for argument. If the main aim is practice in communication skills, it might be possible to add a sequence involving a news conference, public meeting, or parliamentary debate.

However, adaptation of this sort should not be undertaken without a good deal of thought and a reasonable amount of experience with simulations. Really major adaptations should not be attempted: it is better to buy a more suitable simulation.

A special case for adaptation is in the teaching of English as a foreign language. But even here, the teacher should give the simulation an unadapted run first, as it is easy to underestimate the ability and motivation of students involved in a simulation.

Briefing

Briefing is easy, providing it is based on personal participation by the teacher and on careful preparation.

If the students are unused to taking part in simulations, it is useful to spend some time explaining what they are and what they are not. The key points to emphasize are the extent of the powers, duties and responsibilities of the participants, and also the dividing line between reality within the simulation and the fictional background outside. Even if the students have taken part in simulations before, it is advisable to make sure they are aware of these points.

With careful preparation, the organizer will enter the briefing well primed with explanatory notes, diagrams, maps, timetables, deadlines, or whatever else is necessary.

If the simulation is simple, the organizer may be the only person in charge. But with a longer or more complicated simulation it is advisable to have an assistant or even an 'organizing team'. Some teachers who are familiar with simulations prefer giving one or two of the students the opportunity to be organizer as it gives them practice in organization.

Naturally, the briefing should contain no hints or nudges about policy decisions. Therefore, for the organizer, the briefing is a very practical session — a checklist of items to be dealt with, points to be made and queries answered. Should the simulation have several stages, parts of the briefing can be left until later and the information given immediately before the stage in which it is needed. The more thorough the briefing, the less the likelihood of unexpected events arising which could interrupt the simulation or knock it off course.

The organizer should be cautious about giving too much information to the participants during the briefing. A student who is placed in the management team of Blogsville Ropes Ltd might ask, 'Does it manufacture all types of ropes?' If the answers to the questions are available in the documents, it is probably best for the organizer to decline to answer. In most simulations, it is one of the functions of the participants to find out the facts. If the briefing opens the door to this sort of factual question, other participants can start asking similar questions and the organizer can end up by spoonfeeding information to the students. Also, the factual information provided by the organizer in this off-the-cuff manner could be distorted or misleading. It is better to let the students find it out for themselves from the documents.

This will, of course, leave some students dissatisfied and feeling that they do not know enough about the situation. But the organizer can explain that adequate information will be provided when they receive the materials and that the briefing is concerned with the mechanics of the simulation only.

Action

Organizing some simulations is as easy as rolling a ball down a gentle slope. Even with more complicated or lengthy simulations, the organization is simple providing the simulation has been well prepared and briefed. The reason for this is that the action tends to look after itself. It has its own power, its own catalysts and its own initiatives. The problem with a good simulation is not to get it moving but to get it to stop.

In a simulation the organizer is generally in a good position to observe participant behaviour, including fact-learning, strategies, decision-making, problem-solving, and the use of language and communication skills — to say nothing of enjoyment. The organizer should have a notebook handy to jot down various points as they occur — possibly for use during the de-briefing.

The vital task for the organizer during the action is to make absolutely sure that the right materials are available as and when they are required. Once the simulation is under way, the organizer should check the materials, which might have been shown to the students during the briefing, and make sure they are in the right order and grouping for handing out or for availability. In some simulations this is unnecessary as all the materials are given to the participants at the start, but any simulation which involves feeding in materials from time to time requires close monitoring by the organizer.

In addition to the question of materials, the organizer should also prepare in advance for changes in classroom furniture, timetables or role changes which might be part of the structure of the simulation.

With a short simulation — lasting, say, less than an hour — no intervention by the organizer should be required, assuming, of course, that it has been adequately prepared and briefed. But with longer simulations some minor adjustments in the machinery might become desirable or necessary and, very occasionally, a major change is required. In these circumstances, the problem for the organizer is whether or not to intervene and, if so, how and when. Effective intervention may require skill, experience and imagination.

When intervening, the organizer should have two objectives — to interfere as little as possible with the smooth running of the simulation and to select a cover story which fits in with the simulation itself.

Suppose that during a simulation involving interviews or a public inquiry, the organizer notices that the members of the board or panel say, 'Now then, Mr . . . er . . . er'. This may indicate either that the person interviewed does not have a clearly written name tag, or the member of the board has no list of names, or both. At the next suitable break the organizer, adopting the guise of messenger, usher or whatever, provides the necessary name identification material to whoever requires it — perhaps even apologizing and saying that the town clerk's department was responsible for the omission.

With a medium-length simulation, lasting half a day or a day, or a longer simulation which goes on for more than a day, the chances increase of the need for a change in roles. A student may go sick, or may not turn up, thus creating a serious problem. If the simulation consists of teams, the missing member may not make much difference; but if the missing person happens to be the prime minister possessing secret information, it will probably be necessary to halt the simulation in order to make suitable adjustments.

Very occasionally a change of role is necessary because of an internal coup. The prime minister may be overthrown because there was a basic conflict between his 'sell-out' policy and that of public opinion as represented by his country's media. The organizing team may have decided that the prime minister was wrong and should be deposed, or 'forced to resign', or whatever is plausible. Any major change of this nature cannot be dealt with peremptorily. There will almost certainly have to be a break in the simulation until the roles are changed, someone else becomes prime minister and the ex-prime minister takes another role.

A role change would be required in a local affairs simulation if the chairman of the council finished up on the wrong end of a vote of confidence because of repeatedly favouring one group. In this case, the intervention by the organizer might be quite unnecessary, since the participants could deal with the role change within the confines of the action itself — there being no need to hypothesize a general election or military takeover in order to move a person from one chair to another.

It is rarely the case that the unexpected arises unexpectedly. There are usually warning signals and the organizer should watch out for them. In this way, drastic intervention can often be avoided by taking minor remedial action. Even if the major disruption occurs the organizer will have had time to work out some contingency plans.

Suppose the organizer notices that a participant has nothing to do and is looking bored. The question is whether this is slight and temporary or whether something has to be done about it. If something has to be done, it should be within the plausible parameters of the simulation. A note can be handed to the participant from the local council, the editor, the managing director, the shop steward, or the prime minister, asking the participant to help X or Y.

Should any serious misbehaviour occur, probably the best thing is for the organizer to send the person concerned a note asking the participant to come and take a telephone call, or come and see the cabinet secretary, military commander, etc. Having extracted the trouble-maker the organizer can find out what is the matter. It might have nothing to do with the simulation, or it might be a misunderstanding, or a failure of the simulation to allow the participant a full role. Having found out what is the cause of the trouble, the organizer can take whatever action seems appropriate.

It is much easier for the organizer to deal with behavioural problems within a simulation than in normal classroom teaching since there is no escalation of personal antagonism between student and teacher. The organizer, by the nature of the job, is not eyeball to eyeball with the students and consequently is a detached and impartial authority.

The main danger in the action part of a simulation is not misbehaviour so much as inappropriate behaviour, and this may be caused by the failure of the organizer to explain clearly enough what can and cannot be done in a simulation. As has been emphasized earlier, the participants must be told to accept their function: they are businesspeople, town councillors, or world leaders, not magicians, gods or saboteurs.

Supposing, for example, in a history simulation a Saxon king announces that he has moved his army 100 miles overnight. What happens next? If the simulation has been well briefed, then one or more of the other participants will challenge the decision. One challenge would be to point to some of the documents or to common sense and say that 100 miles is an impossible distance to cover in such conditions in such a short time. But a more important challenge would be to say that an order is only an order, and a decision is only a decision, and that the 'facts' outside the room depend on the organizer. Even if it was an order for an overnight journey of only one mile, it would still be up to the organizer to decide whether the order was carried out, whether it was effective, and what other consequences resulted from the order. Assume, however, that the briefing was somewhat inadequate, or that the other participants were not sharp enough to realize that the Saxon king had turned into a Saxon magician, what should the organizer do about the inappropriate behaviour?

One way is to stop the simulation and explain to everybody that this sort of thing is not allowed; but this has the disadvantage of disrupting the flow of the simulation. Another way is for the organizer to send a written message to the Saxon king saying, 'My lord king, your army is foraging for food. It is not possible to march until tomorrow.' This should hold the situation until the next break, during which the organizer can explain the difference between inside decisions and outside facts.

In ordinary teaching it is customary and efficacious for the instructor to step in to correct mistakes of fact. It is also usual to go further than this and offer advice and information about non-factual matters — questions of ethics, values and opinions. For a teacher, these interventions become second nature and habitual. During a simulation there may be an intensely strong and perhaps overwhelming desire to step in, interrupt the action, and convey the correct information or the useful piece of advice. The intellectual justification for this sort of intervention is that if wrong information goes uncorrected it may be learned and reinforced by repetition, and that the organizer should provide the correct facts in order that they may be put to practical use, tried out, tested and learned while participatory interest is high.

Some authors go further than this. They argue that the organizer should ensure that the discussion is relevant, that each person has a fair opportunity to contribute, and that valuable points are discussed in plenary session during the action rather than just within one group.

The trouble with this sort of intervention is that it will kill a simulation stone dead, leaving only an instructor-controlled exercise. There is nothing wrong with instructor-controlled exercises, but they should be advertised as such, not presented under the guise of a simulation, with participant responsibility conferred and then taken away.

On the strongest point of the argument — correct facts — it is necessary to distinguish between the fictitious 'facts' — the Blogsville Company's production costs, the Ruritanian defence treaty, etc — and the non-fictitious facts, the real facts of the outside world which may impinge on the action. In the first case, no 'facts' will have been learned incorrectly; the only incorrect learning is the fictions. As in everyday life people make errors, read documents incorrectly, and so on, and in a simulation they should pay the normal penalties for carelessness, and not have the organizer protect them from the folly of their ways.

On the question of real 'facts' being given incorrectly, these are either in the documents or they are not. If they are in the documents, the author or publisher is in error or the facts are out of date. The organizer should have done something about it before the simulation began. If the 'facts' are not in the documents, they are nothing but allegations made by individual participants and should, as in everyday life, go uncorrected if they pass unnoticed. The organizer can, of course, make a note of the incorrect allegations and point these out in the de-briefing, but he should never interrupt a simulation simply to correct facts. Interventions of this sort disrupt the flow, diminish student responsibility and ownership, constrain behaviour for fear of 'getting it wrong', cause resentment, and open the door to an attitude of 'Now we are back in school again.'

The only case in which intervention regarding 'facts' is justified is when the participants get it wrong in a big way, and are under such a serious delusion that the simulation itself is imperilled. In this case, the best thing is for the organizer to break off the simulation at a convenient point and correct the misunderstanding by whatever method seems the most plausible, depending on the nature of the simulation.

Even when using a simulation to teach English as a second language, the teacher is well advised not to intervene to correct linguistic and grammatical errors. The teacher has to ask the question whether it is to be a simulation or a linguistic exercise, whether the aim is error-free language or successful communication.

In this context Kerr (1977) says:

> In the course of a simulation, the teacher may be tempted to intervene when mistakes are made, or even to introduce brief spells of remedial teaching. In general, this is unsatisfactory from several points of view; the student being corrected finds that his train of thought has been interrupted, while the teacher will probably find that the students are not paying full attention to his explanations, but are anxious to proceed with the simulation. Experience has shown that it is better for the teacher to sit

in the background with a note-pad, jotting down errors as they occur. It is usually convenient to timetable a remedial teaching lesson (immediately after the simulation ends) in which important mistakes can be discussed and remedial practice takes place. Another possibility is to tape record all or part of the simulation and play back the recording to the students afterwards, inviting them to identify their own mistakes as they listen.

This advice about recording a simulation can be useful in contexts other than learning English as a foreign language. Many people speak badly or mumble or fail to put forward their ideas in a way that can be understood. Even reading aloud is a dying art, with the person concentrating so hard on looking at the black print that each word comes out but the meaning stays behind. A tape recording helps to illustrate the point. On the other hand, there are dangers that the intrusion of recording apparatus could inhibit the participants. If tape recorders are to be used, they should be used as often as possible so that the participants get used to them and forget about them.

De-briefing

In the follow-up discussion or de-briefing, the organizer returns to the role of teacher or instructor. The transition need not be abrupt, however, since it may be a good idea to allow one (or two) of the participants to take the chair, particularly if the simulation itself involved this sort of function. In this case, the organizer's contribution would have the same status in procedural terms as that of the students; for example, in giving an account of what the mechanical problems of the simulation were and how the organizer tackled them.

As a general pattern it is useful to go round the table and have the participants explain their own parts in the simulation — what they saw as the nature of the problems and how they dealt with them. Each participant would contribute to the sum of knowledge, which would be particularly useful if there was a divergence in roles and functions. It also adds to the practice in communication skills to be able to explain what one did and why.

The second stage, after everyone has had their say (without comments or discussion), involves general dicussion. The obvious way into this general debate is by looking at the immediate specific questions relating to the outcome — the inquest on the result. But this should not be allowed to degenerate into a rerun of the arguments used within the simulation.

In other words, the de-briefing should move fairly rapidly from the particular to the general. The real value in a simulation will be in the transfer of knowledge and experience to other situations in the future. This is likely to involve general principles — how did the groups organize themselves and was the organization effective? What alternatives were there? Did the group or individuals explore the options, analyse the nature of the situation in which they found themselves, and plan accordingly? How effective was the communication? Were the language and behaviour suitable and appropriate? What lessons did the participants learn? Would they act differently when faced with such a situation in the future?

The organizer may also be keen to get the reactions of the students to the simulation itself — its materials, mechanics and general situation. Here it may be necessary to interpret various remarks. If a student says, 'That was great fun', the comment can mean that the simulation was not much else. If someone says, 'I lost my bit of paper' instead of 'I lost my housing document', this may indicate a lack of realism in the materials. 'I got bored during the last part' could mean that the participant had a part-time or passive role, or that the organizer had not presented the simulation in a satisfactory manner.

Should the students say, 'We'd like to have another go', it could mean that the teacher should have considered the possibilities of a rerun and reached a decision on whether to have one or not or leave it to a vote. Naturally, if there is a rerun, it is advisable to postpone the de-briefing or to curtail it, otherwise too many hints and pieces of advice may be given. But this is a matter for on-the-spot judgement.

In a simulation with a hidden agenda a thorough de-briefing is usually essential. This is particularly true in behavioural simulations. These simulations are likely to stir up emotions which can last much longer than the simulation itself. Some participants may feel exposed or humiliated. Gaining insights into one's character can be an abrasive experience. Some participants might feel that they have not gained insights, but have been cheated or manipulated into expressing attitudes, views or emotions which are contrary to their characters or personalities. Little advice can be given to the organizer in such circumstances, since the problems and likely outcome will already have been anticipated and will probably have been the reason for presenting the simulation in the first place. The organizer should at least make quite sure that there is ample time in the de-briefing to explain why the simulation had a hidden agenda and what it was supposed to reveal.

De-briefings are often missed opportunities. Frequently the de-briefing is:

(a) too brief,
(b) too dictatorial,
(c) too routine and unimaginative, and
(d) follows so closely upon the event that mature reflection is excluded.

As suggested earlier in this section, the format of a de-briefing can involve the participants, and this can be a subject for negotiation and thus regarded as an imaginative extension of the event itself.

For example, it is quite easy to make the de-briefing take the form of a public opinion poll. Each ex-participant is given a clipboard on which a question is written. Their job is to get answers to their own question from as many ex-participants as possible, and provide answers to the questions asked by others. This is not done as a committee meeting with one person speaking at a time but by standing up and moving around from person to person.

The question on each clipboard can be decided by the organizer, or by a committee, or by the pollster. The questions could cover any of the issues already mentioned, either in general terms or related to particular issues:

— How well did the journalists do their job?
— Was the language of the judges suitable for the occasion?
— 'Mistakes are a good thing in simulations.' Do you agree with this statement?
— What advice would you give to the organizer?
— What was your main problem and how did you deal with it?
— Were your space crew colleagues friendly and helpful?
— What was your opinion of the landlord?
— When some participants asked for help, the organizer said, 'I'm not in the tenement (the housing agency, Dart Aviation, Planet Dy, Lingua).' Should help have been given?

When the poll has been completed then various things could happen. The ex-participants could form themselves into groups and collate the results. There could be a general session in which everyone reads out some or all of the answers on their sheets followed by discussions of those answers that were the most interesting, or educational, or revealing. Alternatively, each person could present their own brief research report, outlining their conclusions as well as giving the evidence.

The above example is a de-briefing organized as a research project. It would not be particularly suitable for routine committee simulations where everyone saw and heard everything, but it would fit in with simulations which had hidden agendas or group interactions. Also, a de-briefing in this form adds interest and diversity to the communication skills.

If the simulation was a behavioural one with a high level of personal involvement (as in the case of the simulations described in the chapter on Design) then an immediate de-briefing allows people to get things off their chests, and this is important otherwise dissent and bad temper can spill over inside and outside the classroom. However, there is a case for deferring at least the main part of the de-briefing until another day. This not only allows time for passions to cool and mature reflection to occur, but it also allows a more structured appraisal. With a structured de-briefing there could be written reports or prepared speeches. Different issues could be negotiated or allocated. One person or team could report and comment on what happened in the media group before the broadcast. Other people could look at anxieties, or carry out research into differences and similarities between the event and the real world.

Another virtue of delay is that it can benefit the organizer. Instead of a snap judgement, a delay gives time for a leisurely recall of what was seen and heard and a mental review of what was significant or interesting. Quite often this review in tranquillity turns up interesting aspects, particularly if there is a recording of the event and the organizer can listen or look at what went on. Of course, recordings can take a long time to transcribe and usually this is not worth the effort, but there are occasions when a few minutes of transcribed argument or discussion can be highly illuminating. And if copies are run off then this also makes the students feel that the organizer is really interested in their behaviour. To give just one personal example, a 16-year-old Sudanese girl was the chairperson of a parliamentary committee. She had the job of saying 'Stand up, sit down, shut up.' Listening to the tape revealed that she found 45 different ways of saying this. At first she used the words 'will' and 'would' ('Will/would you please stand up') but later she used 'may' ('You may stand up'). Although 'may' is the politer form, in this case it probably meant 'I have found that you do what I tell you, so I can afford to be polite to you, and my politeness is also a demonstration of my authority.'

Such isolated incidents may seem trivial, but if several of them are mentioned in the de-briefing then they can add up to useful insights into behaviour. After all, the whole point of experiential learning is to learn from experience and modify subsequent behaviour, and small points are often more useful, more personal and more meaningful to individuals than broad judgements.

In some cases the first part of the de-briefing could be replaced by an extension of the original simulation. For example, the occupants of the tenement could report to a plausible body (The Rights of Tenants Association) which could be composed of the former agency employees. The tenants could explain what happened in their individual cases, and the Association could consider whether action should be taken. In SPACE CRASH the group which survived the longest could take on the role of Space School instructors who would hear the reports and the advice from the ghosts of the space crews for the purposes of rewriting the Space Survival Manual.

The importance given to the de-briefing reflects the value of the simulation experience. To allow a simulation to overrun and then have a five-minute question and answer session is to diminish the educational significance of simulations in the eyes of the participants, and also inadvertently to belittle their own efforts. Truncated de-briefings may influence behaviour adversely in the next simulation, leading to less professional attitudes and less personal involvement.

Perhaps the most helpful advice is for the organizer to consider the options, the timing and the format of the de-briefing before running the simulation, preferably in consultation with the participants.

Chapter 6
Assessment

Aims of assessment

Research and evaluation is an area of controversy in simulation
literature, as indeed it is in education generally. The debate is usually
on whether the methodology and design of the experiment is
appropriate and whether the findings justify the conclusions.

But from the point of view of the teacher, there is nothing forbidden,
or even forbidding, about assessment in relation to simulations.

Assessment does not imply some grandiose pie-in-the-sky research
project to test the hypothesis that 'simulations produce greater gains
in critical thinking, decision-making and problem-solving than do other
learning methods', or some such similar generalization. Many authors
have pointed to the unavoidable difficulties of such attempts. Davison
and Gordon (1978) point out that no evaluatory instruments can
readily encompass the many different dimensions of behaviour and
experience involved, and Twelker (1977) emphasizes the problems
caused by the great differences between individual simulations and
between the conditions in which they are used. In the jargon of
research, there are bound to be a great many uncontrolled variables.

The teacher, therefore, should aim only for what is functional and
practical. The idea should simply be to learn something — something
about the individual simulation, the participants, and also the teacher's
own thoughts and behaviour.

It is a pity that virtually all writers on simulations talk about assessment
and evaluation exclusively from the point of view of assessing a
simulation. The only criterion seems to be whether the simulation is
useful, appropriate, stimulating, etc.

For the teacher, this is a restrictive way of thinking about assessment
since it is only half the picture. Just as people can assess simulations,
so can simulations be used to assess people. As well as examining
simulations, simulations can be examinations. Organizations which have
been using simulations the longest — the armed forces and the higher

levels of the civil service — use some simulations for the specific purpose of testing.

At staff colleges, on courses at country houses and in the interiors of ministries and large organizations, simulations are used, together with puzzles, problems, case studies, discussions, etc, as devices not just for training, but also for assessing the participants.

Once the instructor, tutor, constultant or teacher becomes familiar with a particular puzzle, simulation, etc, it can be used to increasing effect for assessment. Practice and experience are needed, as in any other field. There is nothing mysterious about it, nothing really difficult, as every teacher is a professional assessor.

Language and communications

The report of the Bullock Committee, *A Language for Life* (1975), refers to an experiment conducted by the Southern Regional Examinations Board in association with the University of Southampton to study the examining on a large scale of oral English. Some 450 candidates were divided into four groups, each of which took a different form of oral examination:

1. reading a passage and talking with the examiner;
2. making a short speech or lecture and answering questions;
3. talking to the examiner about a diagram previously studied;
4. participation in group discussion.

These four forms of examination were supported by tape recordings and by an assessment of the candidate's spoken English by his teachers.

The conclusions, which were given tentatively, suggested that the most successful and also the most natural method was the second one — making a short speech and answering questions. Method 3 — talking about a diagram — led to the cultivation of 'civilized conversation', while Method 4 measured ability which was unrevealed in the normal classroom situations. The research workers said that the habits of spoken English could be 'sharpened, enriched and disciplined by intellectual and sensitive attention in the classroom and syllabus'.

However, the Bullock report failed to make a widespread impact on education in Britain, a point acknowledged by HM Inspectors in *Bullock Revisited* (1982). The HMIs said that the recommendation that every secondary school should have a 'language across the curriculum' policy may even have been counter-productive because of suggesting something imposed from above. *Bullock Revisited* proposed that departments within schools should be encouraged to review their aims and achievements in teaching their own subjects, including the use of oral skills, writing, note taking and specialist language. The Inspectors said that too many children were not being taught to follow sustained discussion or ask pertinent questions. They added:

> Important though the written word is, most communication takes place in speech; and those who do not listen with attention and cannot speak with clarity, articulateness and confidence are at a disadvantage in almost every aspect of their personal, social and working lives.

It would be surprising if the Inspectors were not feeling highly frustrated by the fact that report after report was receiving lip service and little else. The Inspectors may well have been looking with assiduity for an opportunity to add compulsion to their pleadings. When the Government asked the Inspectorate to merge two secondary school examinations one might be permitted to guess that they approached this task with relish. Certainly the result was an examination quite different in kind. The General Certificate of Secondary Education examination is based on the ideas of the HMIs which teachers had previously ignored – active and genuine experiential learning. In all subjects the GCSE gives teachers the option of course-assessed work as part of the examination marks. This means, in effect, simulations and other non-taught events are part of the examination. And in English there is, for the first time in secondary schools, a compulsory oral component.

Assessing oral skills

Teachers who have not been used to running interactive events in their classroom have viewed the prospect of assessing oral skills in such events with some trepidation. It is one thing to assess a formal standardized oral test where each candidate takes it in turn to give a prepared speech, but quite another to assess the subtle and dynamic oral exchanges which occur in simulations, role play and non-taught exercises.

This anxiety is almost entirely due to unfamiliarity with interactive learning. There is nothing especially difficult about assessing oral skills in such contexts. Here are some of the factors which suggest that anxieties are due to an incorrect view of the situation.

1. Oral assessment is based on common sense and experience. Everyone forms impressions about whether the people they meet speak effectively. It is not difficult to tell whether a person is using language which is appropriate to the occasion.

2. Because simulations are untaught events there is ample opportunity for the organizer to see and hear what goes on. Furthermore, if the de-briefing is delayed for a day or two this allows the organizer to approach the task of observation singlemindedly, without interrupting the observation by mentally working out what instant judgements might be made immediately the event ends.

3. Because the organizer already knows the materials on which the event is based, and may have run the same simulation before, the talk will not come as a complete surprise. This makes it easier to note what is said, and also what is not said.

4. The organizer does not enter a de-briefing unaccompanied. Self-assessments by the ex-participants are part of the evidence. If recordings or transcripts are available from the event then this also can enhance the assessment.

5. Far more published examples of assessing oral skills are available than was the case when the Bullock Report was written.

One of the major influences in oral assessment has been the Assessment of Performance Unit, which is part of the Department of Education and Science. The APU has not only carried out a series of experiments in oral testing, the actual tests used have been highly imaginative. In one test there are two (fictitious) maps, one of which is out of date and was drawn before a motorway was built. Participant A has the out-of-date map and wishes to visit a friend B, who lives at the other side of the motorway. A telephones B who has the up-to-date map. B has to explain to A how to get to the footbridge over the motorway. In this test, A and B sit facing away from each other, or there could be a screen preventing them from seeing the other person's map.

In another test, the participants are each asked to describe a job they know something about, and then without warning they are paired and given the role of a person in the job they had described (journalist, lawyer, sales manager, chef). They have to argue competitively in favour of a pay rise for themselves.

Quite a different type of test involves a list of 12 materials (plastic tray, black card, white card, jug of water, etc) and participants are asked to design an experiment to find out which of four environments woodlice prefer — dry/light, dry/dark, damp/light, damp/dark. The participants do not have to do the experiment, just design it.

Transcripts from the recordings of these and other events, plus the marks awarded, are given in Brooks (1987) and earlier APU publications. Although many of these events are designed for two participants, Brooks makes the point that teachers can adapt the tests for classroom use by building in extra roles. In effect, Brooks is suggesting that teachers can design fully participatory simulations starting from ingenious ideas.

There are now videos for training teachers in assessing oral skills in the GCSE examination which give examples of assessment of largish groups. The London and East Anglian Examinations Group has a video which includes a news conference given by the characters in *Romeo and Juliet*, plus suggested marks for oral skills. As mentioned in Chapter two, the London and East Anglian boards have a formal and externally marked committee test, with each candidate taking it in turn to be in the chair, give a short talk and answer questions. As many marks can be awarded for questions and answers as for the talks themselves. This formal group test is an innovation in secondary schools. A similar test is used in recruitment for the higher ranks of the British civil service, but here the assessors give each candidate a specific issue (documented) which they have to (a) introduce, (b) give a lead in decision-making, and (c) sum up at the end.

Here are two examples from recordings of the awarding of the Linguan Prize for Literature. In both cases the participants were 16 to 17 year olds.

EXAMPLE A

Television presenter: Hello, and welcome back to the presentation of the awards. The judges have read their pieces by the artists and I think they are now ready. Mr M.
Mr M: We'll start in reverse order of the awards, and we can start as soon as possible. In fourth place is 'Poem' by John B. (Applause.) Order. Shh. And in third place is 'The Power of Love' by Jason H. And these next two is a tough decision. Mr R.
Mr R: This is a tough decision because the entries have been very close this year. But after hours of deliberation we've come to the conclusion that we could not split these two entries so we have made this a draw between 'Love is Here' by Nicholas G, which was a good piece of work, a bit long, but a serious piece of work which is very good, and between 'Young Love' by Craig P, which is a fine piece of creative work, so the competition has been announced a draw by these two people. So congratulations. (Loud applause.)
TV interviewer: Congratulations. And now the two judges will read out these two poems. And first 'Young Love' by Craig P . . .

EXAMPLE B

TV presenter: Well, here we are in the Linguan Towers Hotel for the award of the prize for literature. How do you feel about it?
First judge: In general we are not very happy with uh . . . uh . . . (laughter) because most of them were trying to be too sentimental.
TV presenter: But sir, isn't love supposed to be about sentiment?
First judge: Yes, but . . . we were looking for love in the context of human rights, women's liberation, not two lovers or sentimental stuff like children who are collecting flowers.
TV presenter (to second judge): Flowers and children — is that what your colleague meant by sentimental stuff?
Second judge: No . . . um . . . um . . . (laughter). The prize for the most

original ... um ... and have having put a lot of work into it is ... the author is Foster H ... And congratulations (applause).

In example A all three participants could justifiably be given marks in the middle or upper end of the scale. They spoke fluently and appropriately. Had the judges simply said 'We award the prize jointly to Nicholas G and Craig P' then they would have merited only low marks, since the prize giving is enhanced by the explanations, and brevity is inappropriate to the occasion.

In example B the interviewer could be placed in the highest category. He listened to the answers (which not all interviewers do) and was quick to follow them up and ask pertinent questions. The two judges could be given low marks. The first judge had difficulty in putting his thoughts into words. His remarks about human rights etc. had been taken from one of the letters in the *Linguan Times* which had argued against sentimental works, and he was quite unable to defend this position. The second judge was also somewhat inarticulate, but at least he gave a couple of reasons why the prize had been awarded to Foster H.

Interestingly enough, the participants in example A were Post Office cadets who had obtained few qualifications at school. However, this was their seventh simulation at further education college, and their self-confidence had much to do with familiarity. Their first simulation had been as inarticulate as the judges in example B.

By complete contrast, the students in example B were at a highly prestigious private school, and their parents were from the professional classes. However, this was their first simulation. Clearly most of them, but not all, were uneasy, presumably because they discovered that their academic writing abilities which were so effective in most schoolroom tasks were not particularly appropriate in a simulation about literature. Despite what the first judge said about looking for non-sentimental works, they awarded the prize to the only entrant who had produced a poem about love. All the other authors had entered well-written definitions and descriptions of love, including some imaginative comparisons, but they were essays and dictionary entries on love as seen from the outside. The first judge did not seem to realize that he was arguing against what the judges had done, and the second judge (who disagreed with the first judge: 'No ... um ...') missed the opportunity of giving an interesting explanation.

Here is another example of discourse taken from West Germany where SPACE CRASH was being used in the context of teaching English as a foreign language. The students are aged 15 to 16.

Andro: The information is: Dyans are friendly and they will show us the way to the radio station and there we find food and water. But

Dyans are not drinking water — they need only a kind of dry grass and they never move away from grassy areas.
Erid: Yes. Betelg?
Betelg: We are on a flatland...

In this example of foreign language learning, the recording can be used to assess pronunciation, vocabulary and grammar, but it also gives an indication of each participant's ability to use the English language to communicate meaning.

From the purely communicative standpoint Andro's utterance would receive high marks. While 'The information is' may look clumsy as a piece of writing, in the context of the spoken language it shows the value of signposting, of letting the listeners know what is to come next. The information which follows is precise and logically presented, one thing following naturally after another. The friendliness of the Dyans (a highly important piece of information) is followed by what they would do for the space crew — direct them to safety. The fact that the Dyans eat a kind of grass is followed by the information that they never move away from grassy areas.

Erid's two words conveys some deep information. The meaning is possibly something on the lines of 'I am the self-appointed leader of this group because of my superior ability in taking the chair and assessing the situation. Yes, thank you Andro for your helpful contribution. Now then Betelg, let's turn to you and see what information you can give us.'

A full transcript of what happened to the West German space crew on Dy is given in *Simulations in Language Teaching* (Jones 1982).

In assessing oral skills most authorities favour a general impression mark since this allows the event to be judged as a whole. However, there are also plenty of lists of categories which can be born in mind, consciously or subconsciously. For example:

1. The selection of information/ideas.
2. The order and sequence of presentation.
3. The use of language which is appropriate to the occasion.
4. The clarity — enunciating the words, appropriate pace and volume.
5. Fluency and self-confidence.
6. Effective listening and responding to others.

A search through the publications of the examination boards will reveal pages of categories in which the candidates should 'appreciate the need for...', 'understand the developmental structure of...' and 'be aware that...'. But without actual examples these lists can be more intimidating than enlightening. They also give the impression of being absolute standards, yet in certain situations there may be an irreconcilable conflict between categories. If one is requesting help from a stranger then a fluent appeal for assistance (item 5) might be counter-productive or inappropriate (item 3). The ability to listen sympathetically when someone is angry or distressed and to mutter an occasional 'Yes' may well be the most appropriate reaction (item 3), whereas an attempt to interrupt and give an orderly presentation of information/ideas (item 2) could be interpreted as an unfriendly gesture designed to show superiority.

Lists of categories give the impression that they are designed to assess a set speech, not an interactive event. So when assessing the oral skills in simulations a useful starting point could be the four questions mentioned in the chapter on Design: What is the problem? Who are the participants? What do they have to do? What do they do it with? Assessment must depend on the assessor knowing each individual situation, and it really cannot be done by merely looking at the words and consulting a list of categories.

Basing the assessment on what actually happens rather than on predetermined categories opens the door to unexpected findings. Here are three of my own conclusions, which are based on seeing the same simulations being used in some 20 schools and colleges.

In secondary school, West Indians are usually regarded by their teachers as low achievers. My own impressions are that West Indian students are not only as good as white students in simulations, they are better. They appear to be more articulate, more imaginative and more self-confident. (Jones 1986a) I noticed that it was often a West Indian who became spokesperson for a group. In simulations where a written report had to be presented orally, the white children often read aloud their reports rapidly, head down, and stumbling over the words, whereas West Indian children seemed more likely to raise their heads, obtain eye contact with the participants on the other side of the table, and explain and elucidate at a pace suitable for the listeners. These are, of course, only impressions, not measurable results. In some cases it was a white child who was the most articulate and self-confident, but the overall impression is that in general the West Indian children do better, certainly in open-ended and imaginative simulations. The plausible explanation is social not racial — namely that the West Indian community is a highly verbal one and sets a greater value on the spoken word than is the case with the white community. Evidence of this verbal tradition is given in Edwards *The West Indian Language Issue in British Schools* (1979) and is similar to the findings of Labov (1969) in his work with black teenagers in New York.

A second unexpected piece of evidence was that 11-year-olds do better than 14-year-olds in simulations, at least in those cases where the children are unfamiliar with the technique. I had assumed that if

14-year-olds had difficulty in organizing themselves and communcating effectively in a particular simulation, then it would be pointless to try it with 11-year-olds. Quite by chance I was asked if my type of simulations would work with the younger age group, and with considerable misgivings I decided to put the question to the test. The results showed that the younger children worked better together in groups. They consulted each other more effectively. The older children sat in groups but tended to work as individuals. In many cases the language of the 11-year-olds was more articulate and more appropriate to the situation. In BANK FRAUD a couple of 11-year-olds who had the role of executive managers spoke for 3 minutes 50 seconds explaining why they had chosen A and B, and not C and D, for a team to investigate fraud. The 14-year-olds tended to make a one sentence announcement, rather similar to the behaviour of the judges in example B in THE LINGUAN PRIZE FOR LITERATURE. One possible explanation for this is puberty — shyness, lack of confidence, sensitivity, personal insecurity. This seems unlikely to be the cause in view of the ease with which teenagers can change their behaviour after experiencing their first two or three simulations. A much more likely explanation is that the hidden curriculum in most secondary schools teaches children

(a) that the main aim of education is to enable children to produce answers which are brief, factual and quick, and
(b) that a teacher's job is to guide the learners to 'the correct answer'.

If children have had several years of didactic teaching it takes them some time to readjust their expectations, concepts, and behaviour. In their first simulation they keep looking over their shoulders to catch any signals from the 'teacher'. They tend to sit like small birds with open mouths waiting for the grubs to be thrust in.

A teacher who had observed the same simulation being run with different age groups in the school said, 'I was surprised and disappointed that the older children did worse than the younger ones. They just seemed to be looking for quick answers. It must be a result of the teaching methods here.' Another teacher put it even more succinctly when referring to the 14-year-olds — 'We've unabled them.' It would seem that the younger children, because of their experiences in primary school, were more professional in their behaviour, whereas the older children had learned that in secondary school professional behaviour was not required, and that their job was to find brief answers for teachers who would give whatever help was necessary. However, most teenagers learn relatively quickly from simulation experience. They learn that they are required to organize themselves, to co-operate, to think for themselves, to take initiatives, to exercise power. Simulation experience teaches them that asking the right sort of questions is just as important as producing answers. Any shyness tends to be overcome because of the professional duties required of them, because there is no teacher to pick on them when they make mistakes, and because the simulation experience is (usually) both enjoyable and educational.

The third surprising finding relates to life skill courses at the upper end of secondary schools and in further education. I had taken it for granted that courses which included outside visits (factories, law courts) and visits from professionals (magistrates, managers, trade unionists) would reduce anxiety about the world outside the classroom and encourage more positive attitudes. I carried out a small scale study on anxiety levels among young people at a further education college (Jones 1986b, 1987a) in order to see whether experience of simulations diminished anxiety. A test was given before and after an eight-week course during which one afternoon a week was devoted to simulations. This provided evidence that anxieties did diminish. However, a control group was also given the anxiety test before and after their course and this revealed a surprising increase in anxiety. Discussing the matter with the course tutor, and assuming that the findings were not due to chance factors, we finally concluded that what must have happened was that the supposedly active learning was not active, it was passive learning on tour. We further surmised that the control group discovered that the world was more complex than they had at first supposed, and this increased their anxiety. A similar increase could also have occurred in the case of the experimental group, but this was offset by doing simulations which provided the opportunity of finding out that they could cope in a professional environment.

The finding is similar to the results of an American experiment by Lipsky (1981) into anxiety and attitudes towards disabled people. The experimental group used eye patches and performed tactile tasks. They showed a decrease in anxiety and an increase in positive attitudes. However, the control group, who watched a video presentation of 'How do you feel about people with disabilities', showed an increase in anxiety and more negative attitudes. As Lipsky commented, it appeared that the video had reinforced existing fears. However, in the light of the life skills experiment it could be that new anxieties were felt by both groups in the Lipsky experiment, but that in the case of the eye patch group this was counter-balanced by practical experience.

Taken together, the unexpected findings relating to race, age and life skills demonstrate that open-minded assessment of simulations can be highly revealing for both organizer and participants.

Assessing behaviour

Some organizations use simulations and role-play exercises as tests not only of language and communications but also of behaviour. Contact between a member of an organization and someone outside involves behaviour, and the organization is usually very concerned to see that the behaviour is appropriate and effective. In certain jobs and professions, both appointments and promotions depend a great deal on behaviour.

Sometimes the assessment is recorded formally. For example, if the simulation involves an interview of some sort (appointments board, personal interview, media interview) there may be a standard form for the instructor, tutor, consultant, etc to record the result. See below, for example.

Criterion	Standard achieved			Remarks
	Above average	Average	Below average	
Posture and deportment				
Clarity of expression				
Confidence				
etc				

But whether the assessment is formal or informal, the main question to be asked is whether the behaviour is appropriate to the circumstances. This question can be asked about any participant in any simulation. Depending on the participant's function and the specific job and circumstances, the question can be divided up into various parts. For example:

- Ability to make a point
- Ability to stick to the point
- Ability to search for options
- Ability to work out possible consequences
- Degree of courteousness, sympathy, understanding, honesty, diplomacy, etc.

These ideas can form the basis of a teacher's assessment of the behaviour of participants during a simulation.

In-depth interviews

Assessment can be retrospective. As well as observing what happens during a simulation, the teacher can use in-depth interviews after the simulation to assess both the participants and the simulation itself.

This can be regarded as a supplement to the de-briefing. It can cover similar ground, but whereas the de-briefing is often generalized, the interview can get down to the particular and the individual. Since de-briefing sessions often tend to be on the skimpy side, the interview

101

can be a highly valuable tool of assessment and learning. Its value lies in revealing what otherwise might remain unknown or obscure.

In-depth interviewing is rather like beachcombing. All sorts of interesting and curious things come to light , some of them most unexpected. It can reveal insights into attitudes, events and motives which range much wider than the preceding simulation event.

The interviewer will normally be the organizer, but other options could also be considered — a colleague, an observer, an ex-participant, a student from another class, a visitor. Here are a few extracts of comments from children in multiracial classes during interviews with the author.

A 16-year-old Moroccan boy in secondary school after a simulation involving Private Members' bills:

> We've done this and the teacher's out of it, right? And everyone is responsible for their own working. Because, normally, I'm quite good at English, and yet this still increases your knowledge about the world. Like some of the words that we put down here I would never use. I'd never use them, I'd never think of using them. I mean, instead of saying 'offence' I'd say 'a crime'. We would not say 'male' or 'female', we'd say 'boy' or 'girl'. And 'sexual pleasure', we would not even say 'sexual', we'd say 'sex' and that's it. I mean, it's the way we speak. Actually doing this sort of thing we actually learned sort of how law English is.

A 16-year-old Italian girl in secondary school after the simulation TELEVISION CORRESPONDENT:

> Doing the usual lesson is so easy for us because all we have to do is to write it out, and the teacher tells me that this is good, good, good, and I know what to put in it already and it is easy for me to put it on paper and just finish it up. But when I come to this then I quite like it. I am for once understanding what I am doing. I am using difficult words but I understand them. But in a normal lesson, you are just guessing.

A 16-year-old Sudanese girl, also after participating in TELEVISION CORRESPONDENT:

> It teaches you more ... I mean ... if the teacher says to do something, OK you understand me, and you kind of argue, the teacher is bound to tell you the answer because there are so many people in the class, and they just give you the answer and not explain what is going on as they have to go to the next person who needs help. But once you are co-operating all together and sitting down and doing it on your own ... and I find it really interesting, and it's really good, and it's really helpful.

A 17-year-old Jamaican girl in a further education college after participating in THE LINGUAN PRIZE FOR LITERATURE:

> I was the television presenter. I think it was a lovely show, and hope we have another one next year. It was very glamorous and everything went wrong, and we had some lovely poems in, and some very good results.

Some of the points are shrewd. The Italian girl makes a contrast between what is easy and what is understood — most people would assume that these always go together. All the speakers showed natural verbal ability, and the use of repetition in making their points is impressive.

Questionnaires

Questionnaires are probably the most common tool for:

(a) assessing the participants' behaviour/skills/learning;
(b) evaluating and comparing simulations as events.

Usually questionnaires are tailormade by the teacher (or participants, research worker, author) to fit particular events. They can be used for small groups or for much larger experiments. They can be used after the simulation, or both before and after.

There are problems. Bloomer (1974) remarks: 'Questionnaires are prone to the danger that the teacher discovers not what has been learnt, nor even what the pupils thought they had learnt, but only what the teacher would like them to have learnt.'

This touches on one of the main problems — what questions to ask? If the teacher limited the questions to the main objectives for introducing the simulation in the first place, the answers will similarly be limited and the questionnaire will not reveal whether any non-specified objectives were achieved.

It is useful, therefore, for the questionnaire to cover skills as well as facts, emotions as well as subjects, and behaviour as well as learning. Fishing with open-ended questions is revealing: 'One important thing that I found in taking part in the simulaton was . . .'

However, it is preferable to put the open-ended questions first in the questionnaire. If they are added at the end of a list of factual questions, the participant may simply trot out another fact. Here are some sample questions:

The thing that surprised me was . . .
Comment about anything that mattered to you as a person . . .
How did your talking help your thinking?
How did you behave?
Compared with your objectives, did you find what you did satisfactory?
Was the decision-making in your team democratic?
Did you learn anything about being diplomatic?
Would you have liked more time for any of the parts in the simulation?
Did you think it gave you useful practice in (. . . . . .)?
How could you have done better?

Did you consider ethics or only material values?
Would you have introduced the simulation any differently?

If the questionnaire is to be used for statistical purposes, some questions must have quantifiable answers. They can be

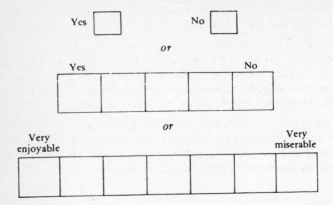

Statistics derived from such questionnaires can be not only misleading, but sometimes the exact opposite of what they are assumed to indicate.

An illustration of this is the figure of 3.6 per cent of participants who listed themselves as being 'totally disinterested' after a large-scale current affairs simulation which lasted for two days at a polytechnic. The implication was that the 3.6 per cent were disinterested in current affairs or the simulation or both. It is the sort of statistic that causes authors to say, 'Yes, we know that there is usually a small percentage of participants who don't like simulations, and here's another example of this.'

Yet on questioning the authors of the research report, it turned out that the participants who had graded themselves as 'totally disinterested' were all members of the team representing the Advisory, Conciliation and Arbitration Service. No other team called them in. They were unemployed. They had sat around for two days doing nothing except watch the other participants enjoy themselves. They were not protesting against participation in the simulation: they were protesting against non-participation.

Questionnaires should not be used in isolation from other assessments and should be supported by an adequate description of what actually happened in the simulation in question.

Two other points can be made. First, it is better to have a questionnaire covering two or more simulations than just one. With one simulation the questions tend to float about in the air. What does 'enjoyable' mean? Enjoyable compared with what? With another simulation, with a lesson in mathematics, with a favourite television programme?

Words like 'useful', 'interesting', 'valuable' derive their meaning from comparisons, and if comparisons are not specified, the whole operation takes on a random flavour with many participants deciding to play cautiously and place the tick in the middle of the range of values.

Secondly, a questionnaire requires to be interpreted in the light of the previous experience of the participants. If they have never taken part in a simulation before their replies to questions may include an extra element of uncertainty, misunderstanding and unfamiliarity because the first simulation is usually the most difficult one.

If the object is to assess simulation X, it is a virtual waste of time to question inexperienced participants. One does not try to assess the value of a specific novel or play by questioning people who have had no previous experience of novel-reading or play-going.

Ethics

Ethical values often find their way into assessments and simulations are no exception. The question of ethics has already been mentioned in relation to behavioural simulations on such controversial issues as race, sex and power. Teachers are understandably anxious that their pupils or students should not be learning the wrong things and acquiring wrong values.

The sharpest attacks on simulations, however, are usually related to competition. There seem to be two main causes of concern about the dangers: one is the desire to win to the exclusion of anything else, and the other is that the values are ofen exclusively materialistic.

Ravensdale (1978) points out that game participants are referred to as 'competitors' and certainly many of the educational games and simulations currently available to teachers are highly competitive. He adds: 'Such games can and do create tensions, show-downs and often fierce competition. That these factors produce a state of adrenalin flow is not surprising; that this extreme attitude can always be useful is questionable.'

Abt (1968) says that 'games of skill have the possible educational disadvantages of discouraging slow learners, dramatizing student inequalities, and feeding the conceit of the skilful.'

Zuckerman and Horn (1973) say that if normal business operations required attention to environmental retrogression, depleting resources and a rapidly deteriorating quality of life, then business simulations would reflect that need. They add:

Paradoxically, if a simulation were designed to include such constraints, it would be a 'bad' training experience, for participants would be trained in methods of thinking and setting decision priorities which would put them at a disadvantage when it was time for them to compete in the real world. However, it would be foolish to condemn the class of business simulations for this significant lack; they are doing their job, which is training.

And Zuckerman and Horn express the hope that some of the simulation designers, particularly colleges and universities, will 'begin to taken an interest in education as well'.

To be fair to simulation designers, there are plenty of simulations about human values and the environment, and to criticize and label other simulations as 'competitive' can be to miss the point. Much of the criticism concerning competition mentions the word 'game'. As Ravensdale says, game participants are 'competitors'. This suggests that much of the criticism results from mislabelling simulations as games, and that the critics are shooting at the wrong target.

DART AVIATION LTD involves business competition. However, the motivation is diverse and to label it as being a competitive simulation is far too simplistic. The aims are to make money, to produce good aircraft, to operate efficiently, to work out the effects of prices on sales, to co-operate with other members of the team, etc. To put this into the same behavioural category as a game of Monopoly is to fail to look inside the event.

TENEMENT is a good example of a caring simulation which can arouse strong passions and denunciations of injustice. The tenants do not compete for benefits, they apply for benefits.

SPACE CRASH might conceivably be regarded as competition against a hostile environment, but this widens the meaning of 'competition' to include almost all human activity. One does not speak of competition in relation to everyday problems – finding a railway station, selecting a library book, writing a poem. SPACE CRASH is nothing if not co-operative.

THE LINGUAN PRIZE FOR LITERATURE has the superficial appearance of a competitive event, but the behaviour is co-operative, creative and professional. The authors engage in authorship, they are not competing against each other like chess players or fooballers. The author who wins the prize does not say 'I've won', but 'Thank you'.

STARPOWER is ethical. It does not teach ethics, but it opens the door to ethical decision-making. It begins as a trading event, but when the favoured group has been granted the power to change the trading rules then it is about human values.

Simulations are the ideal means of introducing a discussion about ethics. Instead of a theoretical debate about what other people should do, the issues are present in personal behaviour. Moreover, the ethics do not have to be a focal point, they often crop up in unexpected remarks, and it is up to the organizer and the participants to decide whether to follow up the incidents.

Questions of morality can be found in most simulations. In THE LINGUAN PRIZE FOR LITERATURE I was once approached by an editor who asked me in confidence, 'Is it permitted to offer a bribe to a judge?' I replied, 'I'm not in Lingua, but if you do offer a bribe then it should be in Linguan money not in British money.' Clearly, the participant was being honest in asking the question. The ethical point for me was whether to mention the incident at the de-briefing. I didn't and the participant did.

In BANK FRAUD it is entirely up the participants to behave as they see fit. As it says in the organizer's notes, 'Lying in a simulation is not the same as lying in a de-briefing, or lying when chatting to friends afterwards.' The point is also made in the notes for participants:

> This is a simulation, and whether you choose to be honest or dishonest is not a question of whether you are normally honest or dishonest. It depends on how you assess the situation as it develops at the Bank, on what you should do in your own best interests as a member of the Bank's staff and also on what you feel you should do as an individual with responsibilities, perhaps family responsibilities.

Some simulations are self-revealing. In the de-briefings of STARPOWER it is not unusual for people to say, 'I was surprised to find myself behaving like that, because I've always said that it was wrong to do that.' This is not a case of people deceiving others, but of discovering that they had been deceiving themselves.

The broad conclusion is that ethics are part of life and therefore are part of simulations, and deserve attention.

References

Abt, C C (1968) Games for Learning. In Boocock, S S and Schild, E O (eds) *Simulation Games in Learning* Sage: California

Bloomer, J (1974) Outsider; pitfalls and payoffs of simulation gaming *SAGSET Journal* 4 3

Brooks, G (1987) *Speaking and Listening; assessment at age 15* NFER-Nelson: Windsor

Bullock, A (1975) *A Language for Life* Report of the Committee of Inquiry appointed by the Secretary of State for Education and Science under the chairmanship of Sir Alan Bullock. HMSO: London

Coleman, B Videorecording and STARPOWER. *SAGSET Journal* 3 1. Also published in Megarry, J (1977) (ed) *Aspects of Simulation and Gaming* Kogan Page: London

Davison, A and Gordon, P (1978) *Games and Simulations in Action* Woburn Press: London

Duke, R D (1979) Nine steps to game design. In *How to Build a Simulation/Game*, the proceedings of the 10th ISAGA conference, Leeuwarden, the Netherlands, vol. 1, 98–112. Also reprinted in Greenblat, C S and Duke, R D (1981) *Principles and Practices of Gaming-Simulation* Sage: California and London

Edwards, V K (1979) *The West Indian Language Issue in British Schools* Routledge & Kegan Paul: London

Elgood, C (1976) *Handbook of Management Games* Gower Press: Farnborough

Greenblat, C S (1981) Seeing forests and trees: Gaming-simulation and contemporary problems of learning and communication. In Greenblat, C S and Duke, R D *Principles and Practices of Gaming-Simulation* Sage: California and London

HM Inspectors of Schools (1982) *Bullock Revisited. A discussion paper.* Department of Education and Science: London

Jones, K (1974) Review of TENEMENT *Games and Puzzles*, June, No. 25

Jones, K (1982) *Simulations in Language Teaching* Cambridge University Press: Cambridge

Jones, K (1984) *Nine Graded Simulations* (SURVIVAL, FRONT PAGE, RADIO COVINGHAM, PROPERTY TRIAL, APPOINTMENTS BOARD, THE DOLPHIN PROJECT, AIRPORT CONTROVERSY, THE AZIM CRISIS, ACTION FOR LIBEL) Max Hueber: Munich. Reprinted under licence under the title *Graded Simulations* (1985) Basil Blackwell: Oxford; now available from Lingual House

Jones, K (1985a) *Designing Your Own Simulations* Methuen: London

Jones, K (1985b) The Linguan prize for literature. Article in *Perspectives on Gaming and Simulation 10, Effective Use of Games and Simulation*, proceedings of the 1984 SAGSET conference. SAGSET: Loughborough

Jones, K (1986a) The survival of the blackest *The Guardian* 25 February

Jones, K (1986b) Simulations and anxiety related to public speaking *Simulation & Games* 17 3

Jones, K (1987a) Life skills, anxiety, and simulations *Simulation/Games for Learning* 17 1

Jones, K (1987b) *Six Simulations* (SPACE CRASH, MASS MEETING, THE RAG TRADE, BANK FRAUD, TELEVISION CORRESPONDENT, THE LINGUAN PRIZE FOR LITERATURE) Basil Blackwell: Oxford

Kerr, J Y K (1977) Games and simulations in English-language teaching. In *Games, Simulations and Role-playing* British Council: London

Labov, W (1969) *The Logic of Non-Standard English* Center for Applied Linguistics: Washington DC. Also reprinted in Labov, W (1972) *Language in the Inner City* University of Pennsylvania Press: Philadelphia, also published by Basil Blackwell: Oxford (1977).

Lipsky, D K (1981) The modification of students' attitudes towards disabled persons. Paper presented at the Annual Meeting of the American Educational Research Association, April 13—17, Los Angeles, California

Masterman, L (1980) *Teaching about Television* Macmillan: London

Masterman, L (1985) *Teaching the Media* Comedia: London

Moses, J L (1977) The assessment center method. In Moses, J L and Byham, W C (eds) *Applying the Assessment Center Method* Pergamon Press: Oxford

OSS (1948) *Assessment of Men. Selection of personnel for the Office of Strategic Services* Rinehart: New York

Ravensdale (1978) The dangers of competition *SAGSET Journal* 8 3

Shelter (1972) *TENEMENT* Shelter: London

Shirts, R G (1969) *STARPOWER* Simile II: Del Mar, California

Shirts, R G (1970) Games people play *Saturday Review* 16 May

Shirts, R G (1975) Ten mistakes commonly made by persons designing educational simulations and games *SAGSET Journal* 5 4

Shirts, R G (1977) *BAFA BAFA* Simile II: Del Mar, California

Stenhouse, L (1975) *An Introduction to Curriculum Research and Development* Heinemann: London

Thatcher, D (1986) *Introduction to Games and Simulations* SAGSET: Loughborough

Townsend, C (1978) *Five Simple Business Games* (GORGEOUS GATEAUX LTD, FRESH OVEN PIES LTD, DART AVIATION LTD, THE ISLAND GAME, THE REPUBLIC GAME) CRAC/Hobsons Press: Cambridge

Twelker P A (1977) Some reflections on the innovation of simulation and gaming. In Megarry, J (ed) *Aspects of Simulation and Gaming* Kogan Page: London

Wittgenstein, L (1968) *Philosophical Investigations* Basil Blackwell: Oxford

Zuckerman, D W and Horn, R E (1973) (eds) *The Guide to Simulations/Games for Education and Training* Information Resources Inc: Lexington. (The editors of the latest edition (1980) are Horn, R E and Cleaves, A.)

Societies and journals

SAGSET: The Society for the Advancement of Games and Simulations in Education and Training. The Society's journal, *Simulation/Games for Learning*, is published quarterly. The proceedings of SAGSET's annual conference, which is usually a series of workshops, are published under the general title *Perspectives on Gaming and Simulation*. Details from The Secretary, SAGSET, Centre for Extension Studies, University of Technology, Loughborough, Leics LE11 3TU.

Simulation & Games is the official publication of the North American Simulation and Gaming Association (NASAGA), the International Simulation and Gaming Association (ISAGA) and the Association for Business Simulation and Experiential Learning (ABSEL). It is a quarterly, and is available from Sage Publications, Beverly Hills, California, and London.

ISAGA newsletter: Available to ISAGA members only, it usually contains information about courses, conferences and publications, plus occasional articles.

Simjeux/Simgames: A Canadian quarterly available from Pierre Corbeil, 690, 104c Avenue, Drummondville, Quebec J2B 49P.